Translating Audio Description Scripts

Text – Meaning – Context:
Cracow Studies in English Language, Literature and Culture

Edited by
Elżbieta Chrzanowska-Kluczewska
Władysław Witalisz

Volume 12

PETER LANG
EDITION

Anna Jankowska

Translating Audio Description Scripts

Translation as a New Strategy of Creating Audio Description

Translated by Anna Mrzygłodzka and Anna Chociej

PETER LANG
EDITION

Bibliographic Information published by the Deutsche Nationalbibliothek
The Deutsche Nationalbibliothek lists this publicationin the Deutsche Nationalbibliografie; detailed bibliographic data is available in the internet at http://dnb.d-nb.de.

This publication was financially supported by the Jagiellonen University in Cracow.

Library of Congress Cataloging-in-Publication Data
Jankowska, Anna, 1980-
Translating audio description scripts : translation as a new strategy of creating audio description / Anna Jankowska ; Translated by Anna Mrzygłodzka and Anna Chociej. – Peter Lang edition.
    pages cm. – (Text - Meaning - Context: Cracow Studies in English language, literature and culture; Volume 12)
    ISBN 978-3-631-65344-9 (Print) – ISBN 978-3-653-04534-5 (E-Book)
1. Translating and interpreting–Technological innovations. 2. Multimedia systems–Research. 3. Audio-visual equipment–Technological innovations. 4. Translating and interpreting–Technological innovations. I. Mrzygłodzka, Anna, 1986- translator. II. Chociej, Anna, 1987- translator. III. Title.
    P306.93.J36 2015
    418'.02--dc23
                                    2014047053

ISSN 2191-1894
ISBN 978-3-631-65344-9 (Print)
E-ISBN 978-3-653-04534-5 (E-Book)
DOI 10.3726/978-3-653-04534-5

© Peter Lang GmbH
Internationaler Verlag der Wissenschaften
Frankfurt am Main 2015
All rights reserved.

Science should not contribute to science.
It should contribute to the good of men.
/Aaron Ciechanover/

I would like to thank Professor Elżbieta Tabakowska, Professor Władysław Miodunka and My Parents for the fact that they were and they still are my mentors and role-models in every and true sense of each of these words. I thank Krzysztof Gurszyński, Agnieszka Szarkowska and Agata Psiuk for friendship, constant support and inspiration.

Table of Contents

Introduction 9

I. Part One...15
1.   Audio description in theory... 15
1.1 An attempt to define audio description...16
1.2 Audio description's place in translation studies 19
1.3 The history and the current state of audio description.................25
2.   Audio description in practice 39
2.1 Audio description creation..39
2.2 Audio description beneficiaries..51
2.3 The importance of audio description 54

II. Part Two...57
3.   State of the art 57
4.   Research methodology 63
5.   Time-consumption analysis... 63
5.1 Research procedure..64
5.2 Results of the time-consumption analysis 67
5.3 Time-consumption analysis: summary and conclusions 70
6.   Pilot study.. 72
6.1 Research procedure..72
6.2 The results of the experimental group research.............................78
6.3 Control group survey results 89
6.4 Summary of the results and conclusions..91
7.   Comparative analysis of the scripts... 94
7.1 Selection of the research tool 95
7.2 Research procedure.. 101
7.3 Detailed analysis.. 102
7.4 Comprehensive analysis 111
7.5 Comparative analysis of the scripts – conclusions 113

III. Final Conclusions.. 117

References... 121

Appendix 1 129

Appendix 2 133

Introduction

Audio description (AD) is an additional verbal description of the most important elements of a picture, i.e. a technique that provides visually impaired people with information that other viewers perceive only visually. This way, audio description gives the visually impaired access to visual and audiovisual products, in a broad sense of this word. The technique of audio description is used in television, cinema, during stage and sports shows, as well as in museums and galleries. Also DVD and Blu-ray recordings are made available through audio description. It is commonly said that it is a young, practically fledgling field. In fact, audio description has existed for almost thirty years, even for over fifty, according to some, and it was invented as a practical service that provides visually impaired people with an access to visual and audiovisual culture. Initially, it was mostly of interest to its beneficiaries: visually impaired people and non-governmental organisations that act in their interest and spend the funds they gather on, e.g. audio describing films and theatre shows. In the course of time, however, audio description was found interesting by scientists – universities conduct research, organise scientific conferences and teach the difficult craft of audio describing. Recently, many national and international initiatives were undertaken, aiming at the development of audio description and integrating theory with practice. With such a dynamic development of audio description in mind, in March 2013 the participants of Advanced Research Seminar on Audio Description (ARSAD), which takes place in Barcelona every two years since 2007, decided unanimously that it cannot be called a field in its beginning phase any more. Apparently, the initiatives of the groups of theoreticians and practitioners are followed by legislators as well. Already in 2007, European Parliament issued the Audiovisual Directive, whose main assumption is to provide elderly and disabled people with an opportunity to participate in social and cultural life through, e.g., the implementation of audio description (European Parliament and Council 2007).

Audio description has been present in Poland since at least 2006. It has been promoted by non-governmental organisations that undertake many initiatives whose aim is to provide the visually impaired with an access to visual and audiovisual culture. Collaboration with cinemas, film festivals, public television, theatres, museums and football clubs was a part of these initiatives. Audio description also became a subject of interest for Polish researchers from a number of academic centres – the scope of this interest is best proved by the fact that the group of Polish scientists was the largest national representation during the above mentioned ARSAD conference.

In one of the first articles on audio description in Poland, three basic problems concerning the implementation of audio description in this country were named: the lack of legal regulations and financing mechanisms, the need of audio description standards, and the absence of experienced, or trained audio describers (Jankowska 2008: 244). When taking into consideration the fact that the first Polish cinema screening with audio description took place over eight years ago, and the article quoted in this book was published only six years ago, it is hard not to appreciate how much Poland has achieved in the field of audio description implementation until today. In this short period the attempts to develop certain legal solutions that would regulate the question of implementing audio description in Polish media ended up successfully. Since the 1st of July 2011, in accordance with the Amendment to the Broadcasting Act, Polish broadcasters are obliged to introduce improvements for visually and hearing impaired people, and they should constitute at least 10 per cent of the programming, excluding commercials. The new law somehow solved the problem of financing audio description, and its presence – at least in television – is no longer the effect of the good will of broadcasters and the effort of a group of enthusiasts, but became a regular service, which is paid by the broadcasters just like any other audiovisual translation. A change in the field of standards also occurred – at the moment of writing this book as many as two documents on the rules of audio description creation were available, and a book which presents the rules of audiovisual translation, including audio description, and a textbook on writing audio description was soon to be printed. One can also find courses, which usually take one day and are organised by audio description foundations, as well as courses which are a part of MA or postgraduate studies.

Although we appreciate all those initiatives, it is hard not to notice that the steps taken to solve the above mentioned problems are insufficient and the situation connected with the availability of cultural properties for the visually impaired in Poland is still unsatisfying.

The groups lobbying for the availability of visual and audiovisual culture for visually impaired people had high hopes for the legal regulation of audio description. They expected that a legislative regulation would solve the problem of the profitability of audio description, whose presence should not depend on the number of people who use it (Szymańska in Jankowska 2008: 244). The above mentioned amendment met the expectations to a certain degree. However, one should notice that only television broadcasters are obliged to implement audio description, and there is still no obligation to audio describe films screened in the cinemas or those distributed in a digital formats. As an effect, cinema screenings, or even the issues of DVD and Blu-ray formats are still financed mostly by non-governmental organisations for visually impaired people. The question of audio description in television also leaves a lot to be desired. From the perspective of broadcasters, the implementation of the new law is connected with high costs. Audio description cannot be considered a profitable product – its relatively small group of recipients is not a target group for advertisers. When trying to save money, the broadcasters apply various practices. The first of them is such an interpretation of the Act's regulations that is beneficial for the broadcasters. According to their interpretation, the features for both visually and hearing impaired people must constitute at least 10 per cent of the quarterly time of the programming broadcast, and the proportions between them depend on the broadcaster. As an effect, broadcasters use the facilities for hearing impaired people[1] more often, since they are cheaper than audio description. Another practice of this kind is the qualification of sports broadcasts with commentaries as audio described programmes. At this point it is worth mentioning that a classic sports commentary does not fulfil the requirements of the visually impaired and differs significantly from the audio description of sports events. Yet another practice of this kind is adding audio description to programmes, when the cost of its creation is not big when compared to the length of the programme. A perfect example of such practice is audio description prepared for talk-shows. As the name suggests, productions of this type mostly consist of conversation, which, in accordance with the programme's convention, is held in unchanging place and time. Only guests change, and their entrance is announced by the host and acclamation of the audience, and their physical appearance, although it is quite often characteristic, is not necessary to understand the content and the sense of the conversation. Therefore the sense of adding audio description to such a kind of programme is questionable. However, from the purely financial perspective such a procedure is beneficial – the creation of audio description which consists of several one-sentence lines that inform about the physical appearance of a next guest does not require hiring a professional audio describer. Usually this job is done by television crew members.

From the perspective of time, the question of a real usefulness of the rules and standards of audio description creation, at least in its present form, is another aspect that raises doubts. The rules and standards of audio description that

---

1 Subtitling for the deaf and hard of hearing, Polish Sign Language) or Signed Polish.

function on the Polish market are partially based on foreign standards, which are considered to be tested, and on their authors' experience. However, it is being asked more and more frequently: what are the bases of the British and American standards, which are used in other countries that considered them as tested? Has anybody ever asked visually impaired people about their opinion on the solutions suggested by those standards? Are they something more than a collection of anecdotic commandments and a description of the already applied practices (Udo & Fels 2011 and Fryer 2009)?

The question of the training and experience of audio describers is also not satisfactory. The courses and trainings which are offered currently are short and cover no more than 12–16 teaching hours. In such a short period one can gain some basic skills, but there is practically no way of developing them, since the market of commissions for audio describers is too small and dominated by a several professional audio describers, who mostly learned the rules of audio description themselves.

The aim of this book is to suggest a solution that will help to decrease costs and time of audio description production and at the same time will allow to preserve its high quality. The solution we propose is the translation of audio description scripts, written by experienced British specialists, from English to Polish in order to add them to dubbed films screened in Poland. The dubbed versions were chosen deliberately, since at the time when this research was conducted, audio description in Poland was most frequently added to Polish films or the ones dubbed into Polish.

There are a couple of reasons which let us assume that translation of audio description scripts from English to Polish is not only possible, but also beneficial when it comes to time consumption, costs and quality.

Firstly, in the multiple-stage process of creating audio description script the most difficult and time-consuming stage is the one connected with the decision of what should be described and how (Remael & Vercauteren 2007). Realisation of this stage demands watching a given audiovisual material many times and deciding which visual content should be conveyed through words. Then, a script prepared by an audio describer is consulted with a blind person in order to make sure that the description is enough for the target audience to follow the plot of a film or a programme. The solution we propose assumes elimination or limitation of the above mentioned stages. A translator who is given an audio description script which was written by a British audio describer and consulted by a blind person, does not need to watch the film many times, make a decision concerning the selection of the content and consult it with a blind person. It is enough to translate the script and then synchronise it with the space available between dialogues. It seems that in this way the time consumption of audio description scripts creation can be decreased.

Secondly, in accordance with the present rates[2] on the audio description market, the cost of audio description prepared for one act[3] of a film amounts to the gross value of 100–120 PLN, whereas audiovisual translation of one film act amounts to the gross value of 50–80 PLN. Therefore, one can assume that if the creation of Polish script is based on its translation, and not writing it from scratch, it would be possible to decrease the cost of audio description script creation.

Thirdly and most importantly – when it comes to the quality of the scripts translated from English – the United Kingdom is a European leader in the field of the implementation of audio description, and the British film audio description tradition dates back to the 1980s. This is why there are many experienced audio describers in this country, who are professionally trained or have an opportunity to graduate in accessibility postgraduate studies before they start their work. There are also many possibilities to develop and improve the acquired skills in practice. Therefore, one can assume that the quality of audio description scripts created by British professionals is high, and a translation of the scripts to Polish should preserve the quality comparable to or higher than the quality of scripts written by Polish audio describers.

Fourthly, translation of scripts is a common practice applied by international companies that provide, e.g., subtitles for DVD. Instead of asking several translators of different languages to prepare a dialogue list based on a film and a scenario, they are usually provided with a so called "template". It is a file of English subtitles with set time codes, which is then translated by the translators into their native languages. It seems that this practice can be used in the field of audio description.

The book has **three parts**. In the **first part** the idea of audio description is presented in both its **theoretical** and **practical** dimension. The **second part** of the book presents the results of a study, which was conducted to check whether the translation of audio description scripts from English to Polish is a solution that is less time and money consuming, and whether the scripts created through application of the translation strategy preserve the quality that satisfies the target

---

2   Data from the Sevents Sense Foundation.
3   An act is a unit of pricing for audiovisual translations in Poland, meaning every 10 minutes of audiovisual material.

recipients. The **third part** of the book includes final conclusions and the suggestions concerning further research.

Part one of this book is a kind of an introduction which is necessary to understand the sense, aim and the way of conducting the research. The **first chapter** of this part investigates theoretical aspects of audio description, such as the question of terminology applied to name the phenomena of audio description, as well as many of its definitions formulated from the perspective of disability studies, text linguistics and social sciences. Then we tried to place audio description in the field of translation studies, particularly in the frames of audiovisual translation and Jakobson's categories of intersemiotic, interlingual and intralingual translation. The **second chapter** presents the history of the beginning and development of audio description. Due to the character of this book the history of audio description was limited to the very beginnings of the audio description idea's functioning, and the past and present situation in the United Kingdom and Poland. This knowledge seems to be necessary to understand why the scripts from United Kingdom are the ones to be translated into Polish and to be used in Poland. In the **third chapter** of the theoretical part you can find the practical aspects of audio description creation, like the process of its creation and the strategies applied to create a script. This chapter is also focused on the beneficiaries of audio description and the sense of its creation. **Part two of this book** presents the results of the study that was conducted to check whether the translation of audio description scripts from English to Polish can be a less time and money consuming solution than their original production, when preserving high quality of audio description. The study was comprised of three experiments: the first one was a **time-consumption analysis**, the second one – a **pilot study**, and the third one – a **cognitive comparative analysis of the scripts**. Each of the experiments was presented in a separate chapter including subchapters that describe the procedure and research methodology, and present and summarise the results of the experiment. Each chapter includes final conclusions. General conclusions are presented in a separate, **third part** of the book.

I. Part One

The first part is divided into two main subchapters, in which we present both **theoretical** and **practical** attitude to audio description. The chapter devoted to an investigation of the **theoretical** aspects of audio description presents an attempt to name it, define it and place it in the frame of Translation Studies. It also presents the history of the beginnings of audio description, as well as its tradition and the state of its development. The chapter which is focused on **practical** aspects investigates the strategies and the process of audio description creation.

1. Audio Description In Theory

The following chapter, which investigates **theoretical aspects** of audio description, is divided into three subchapters. The first of them is devoted to the issues of the **nomenclature and definition of audio description**, which, although it is considered to be a young field, has many descriptions and definitions. The subchapter includes an analysis of the names which are used to define the phenomenon which is most commonly called "audio description", its definitions formulated by active audio describers, its target group, legislative organs and scientists, who treat audio description as a subject of investigation of such fields as disability studies, sociology and linguistics.

Although it may seem surprising, audio description is also a field investigated by Translation Studies (TS) researchers. The second subchapter analyses audio description from the perspective of TS and is focused on such issues as the place of audio description in translation studies, particularly in the field of audiovisual translation, as well as Jakobson's categories of intersemiotic, interlingual and intralingual translation. Such a detailed investigation of the approach of TS seems to be necessary, not only due to the character and aim of this book, but also because the perception of audio description in the frames of translation studies has measurable practical consequences.

The third subchapter describes briefly the **history and the current state of development** of audio description. It is focused on the situation in Poland and in the United Kingdom, because the knowledge of this scope seems to be necessary to understand the sense and the aim of this research and the reason why the British scripts are those to be translated into Polish and used in Poland.

1.1 An attempt to define audio description

"The art of speaking with pictures" (Orero 2007), "a picture painted with words", "audio description", "video description" (Cronin, Evans & Pfanstiehl 1997), "description service", "art", "a type of poetry" (Snyder 2007:100), "audio narration". These are only some of the definitions of audio description, and although it is a relatively young field, it has a great number of descriptions and definitions. What is more, the definitions of audio description are subject to multiple modifications, since its scope of influence increases, which is the effect of the dynamic development of audiovisual media and modern technologies. Therefore we deal with the definitions of audio description formulated by its visually impaired recipients, state and international institutions that legitimise its implementation, active audio describers and scientists, who define audio description from the perspective of such fields as disability studies, sociology, linguistics and, which may seem surprising, translation studies.

1.1.1 Audio description or maybe...

Except for the phrases full of emphasis, like "the art of speaking with pictures" or "painting with words", there are also some more technical terms. Currently, the most frequently used term is "audio description", but in English literature one can find other terms, which are used simultaneously, like "video description", "audio vision", "descriptive narration", or "audio narration" (Hernàndez & Mendi-luce 2009: 162). Not all of these terms are used in Poland, where the most frequently used names are "audio description", "audio narration", "narrative description" or "audio commentary". The multiplicity of terms used to name this relatively new field proves its dynamic and constant development. Below you can find a brief presentation of the questions connected with its proper nomenclature.

The terms frequently used in Poland are "audio description" and "audio narration". Italian researcher Saviera Arma (2012: 15–20) postulates that the terms "narration" and "description" are too different to use them interchangeably. Firstly, narration is almost exclusively concentrated on a description of action, while audio description describes also protagonists, elements of scenery, sounds, colours etc. Secondly, audio description's aim is not to tell a story, but to describe visual elements which are necessary to understand the plot. Meanwhile, narration not only tells a story, but also has a particular point of view. Thirdly, when narrating, the first or third person is used mostly, as well as various tenses. Audio description uses mostly the third person singular and a present tense. Concerning all of the arguments mentioned above, an "audio narration" should be rather a combination of an original sound track (i.e. dialogues, sound effects and music) with a narration with sound. The final effect of such a combination would be similar to an audio book or a radio theatre of imagination known from the Polish Radio. A similar situation is related to a frequently used term of "narrative description". Accordingly, Saviera Arma has rightly stated that the characteristic features of narration do not suit the current standards and aims of audio description that postulate absolute objectivity and impersonality. Therefore it seems that terms including the word "narration" are not recommended to be used in order to describe the technique that makes the visual and audiovisual works available to visually impaired people.

Arma (2012: 22) claims that using the term "audio commentary" is also incorrect, since this name is reserved for an additional audio track with commentaries and explanations to a film, which are read by one or more authors, and whose aim is to convey more information than the picture, or to make the audience laugh. Therefore, the aim of audio commentary is not to describe verbally the content presented in the picture to visually impaired people, but to provide the audience with additional information concerning the cast or drawing attention to the non-obvious facts.

1.1.2 Definition of audio description

As it will be further described in this book, audio description was invented for the purpose of cinema, television and theatre productions. It is also most frequently used in those fields. Probably this is the reason why most of the currently used definitions of audio description are focused on audiovisual and stage productions. However, there are cases of using audio description not only in order to facilitate an access to stage performances or audiovisual productions, but also to national natural heritage, historical and cultural heritage and other social and cultural events (Igareda 2011: 223). Therefore it seems that at this moment the most comprehensive definition of audio description is such a description:

> A set of techniques and abilities whose main objective is to compensate for the lack of perception of the visual component in any audiovisual message by providing suitable sound information which translates or explains in such a way that the visually impaired perceive the message as a harmonious work which is as similar as possible to that which is perceived by the sighted (AENOR 2005 in Igareda 2011: 223).

In other words, it is a technique that provides blind or partially sighted people with those pieces of information, which other viewers receive only visually (Jankowska 2008: 228).

Szarkowska (2011:143) claims that similarly to translation, audio description can be perceived both as a product and a process. The former is an additional audio track including a verbal description of the visual layer, while the latter is the process of its creation.

From the perspective of text linguistics, audio description is sometimes described as a separate type of a text, an audiovisual text (Braun 2008: 17), which is a text that constructs sense on the basis of two communication channels: acoustic and visual one (Chaume 2004: 31) and four types of signs (Delabastita 1989: 199): verbal sound signs (e.g. dialogues and song lyrics), non-verbal sound signs (e.g. music, sound effects, non-verbal sounds), visual verbal signs (e.g. captions and signboards), visual non-verbal signs (e.g. picture composition, special effects, montage). Of course, there are also researchers who claim that audio description cannot be a separate type of text, because when played without a film it is obscure, and therefore it does not meet the requirements formulated by Halliday (Bourne & Jimenez Hurtado 2007: 176). However, it is emphasised that it fits perfectly into the category of multimedia texts formulated by Reiss (Braun 2008: 17), since:

> They are texts which are only part of a larger whole and are phrased with a view to, and in consideration of, the "additional information" supplied by [another] sign system [...] and additionally these are texts "which, though put down in writing, are presented orally" (Reiss 1981: 126 za Braun 2008: 17).

Sometimes audio description is also defined in relation to the art of rhetoric. Braun (2008: 17) suggests that audio description can be compared to a rhetorical figure, hypotyposis, which is a verbal description of a visual scene, or a creation of a picture through words (Braun 2008: 17 in Eco 2003). Meanwhile, Grodecka (2010: 5) investigates audio description in the frames of aesthetic ekphrasis, which allows a blind person to have contact with a work of art, or a subjective form of writing down perceptual experiences.

To sum up – it seems that currently audio description can be defined in a most simple, and at the same time most comprehensive way, as a technique which provides blind or partially sighted people with pieces of information, which other viewers receive only visually (Institute of Informatics of the Jagiellonian University 2002), thanks to which those people get access to audiovisual and visual products in a broad sense of these terms (Jankowska 2008: 228). At this point, it is also worth mentioning the social dimension of audio description, whose aim – except for enabling an access to cultural products – is also social integration (Jankowska 2008: 232). Such a definition seems to cover all of the fields in which audio description is currently applied – it should be noticed that audio description is used also in museums, when sightseeing around cities, during sports events, concerts, dance shows, to describe pictures in newspapers and books, and even during fashion shows. The definition formulated this way is also open for new media and new ideas of the use of audio description, which will certainly appear soon.

Nevertheless, for the purpose of this book we will keep to the original definition of audio description, described as an additional audio track, which is read by a narrator during the breaks between dialogues, and whose aim is to describe briefly what is visible on the screen (Szarkowska 2008:22).

1.2 Audio description's place in translation studies

With the special character of audio description in mind, it may seem that this field should be developed by specialists in disability studies, psychology, film studies, theatre studies, history of art, etc. This is the case in the USA and Canada (Udo 2009), and, according to the references, at the beginning this was also the case in Europe[4]. However, in the course of time, audio description generated interest of translation studies researchers, who saw many common features of translation and audio description. Thanks to a particular activity of the scientists gathered in research groups that deal with audiovisual translation – like Transmedia[5], Transmedia Catalonia[6] and AVTLab[7] – audio description seems to be currently a domain of translation studies researchers, at least in Europe.

From the perspective of translation studies, audio description is counted as audiovisual translation, but it is also perceived as a perfect example of intersemiotic translation understood in accordance with the definition proposed by Jakobson (Orero 2005, Benecke 2007a, Bourne & Jimenez Hurtado 2007 and Jankowska 2008). However, it seems that in some cases audio description can be classified in the two remaining categories defined by the Russian linguist, namely interlingual and intralingual translation (Bourne and Jimenez Hurtado 2007 and Psiuk 2010).

---

4   E.g. in 2002 Cinzia Antifona form the Department of Sociology at La Sapienza University in Rome wrote a dissertation on audio description as an instrument that socially integrates blind people.

5   A European research group that gathers scientists and representatives of audiovisual translation business from Belgium, Germany, Portugal and the United Kingdom.

6   A research group at the Autonomous University in Barcelona, whose members are the workers and PhD students of this university, and co-workers from other scientific units in Europe and the world.

7   Research group Audiovisual Translation Research Lab which is a part of the Institute of Applied Linguistic at Warsaw University.

1.2.1 Audio description as audiovisual translation

One of the first scientists who considered audio description as a part of audio-visual translation was Yves Gambier, who classified it, together with translation of scenarios and subtitling for the deaf and hard of hearing as "a challenging type of AVT" and defined it as:

> A kind of double dubbing in interlingual transfer for the blind and visually impaired: it involves the reading of information describing what is happening on the screen (action, body language, facial expressions, costumes, etc.), which is added to the soundtrack of the dubbing of the dialogue, with no interference from sound and music effects. Audio description can be intralingual. (Gambier 2003: 174–176).

The definition proposed by Gambier is still up to date, although one should extend it to voiced-over films and films with so called audio subtitling (Jankowska & Szarkowska 2011). Audio description is added more and more often to those films and it enables visually impaired people who live in countries, in which films are not dubbed, to watch foreign productions.

Audio description can be considered as a type of audiovisual translation, since it meets three of many requirements that determine audiovisual translation. Firstly, – its aim is to make the national or foreign audiovisual products available to those people, who would not understand them without such aid (Diaz Cintas in Szarkowska 2011: 143). This aim is exactly the same as the aim of, e.g., subtitled, voiced-over and dubbed films, but audio description has a different target audience. Secondly – both the source text, on which audio description is based, and the target text are so-called audiovisual or multimodal texts (Braun 2008:17), since they construct their sense on the basis of various meaning codes conveyed by visual and auditive chanels (Jankowska 2008: 229). Thirdly – as it is emphasised by Bourne and Jimenez Hurtado (2007: 176), audio description has a lot in common with many types of audiovisual translation. Similarly to other types of audiovisual translation, the final shape of a target text is determined by restrictions connected with limited amount of available space and time (Bourne & Jimenez Hurtado 2007: 176). However, whereas in some types of audiovisual translation, like dubbing, subtitles or a voice-over version, the target text has to be synchronised with the time of utterances, audio description should fit in the brakes between them.

1.2.2 Audio description and Roman Jakobson's typology

Roman Jakobson (2012: 127, published originally in 1959) differentiated three types of translation. The first one is the **interlingual** translation, which he defined as an interpretation of verbal signs by the use of other verbal signs of the same language. The second type is the **intralingual** translation, which is the proper one, and is based on an interpretation of verbal signs of one language by means of verbal signs of a different language. The last of the types of translation he mentions is the **intersemiotic translation**, which Jakobson also called trans-mutation, and he defined it as a translation based on an interpretation of verbal signs by means of non-verbal signs. Tomaszkiewicz (2006: 65) underlines the fact that Jakobson himself did not comment upon this definition broadly and only a few works on transmutation have been published until today.

Therefore, when adopting Jakobson's *sensu stricto* definition, it is impossible to speak of intersemiotic translation. However, intersemiotic translation can be also understood in a slightly broader context: as a translation of the meaning of a message formulated in one semiotic system in such a way that, thanks to the selection of the most proper signs from other signs system and its most proper combination, results in a message, whose meaning is the same as the meanings of the message that is being translated (Hopfinger 1974: 82). Thanks to such a definition, we can not only talk about **intersemiotic translation** when it comes to a film adaptation of a novel, but also in case of audio description, which trans-lates a picture by means of verbal signs (Bourne & Jimenez Hurtado 2007 and Jankowska 2008).

However, the creation of audio description by means of writing a script which is a verbal realisation of a picture is only one of the ways of how audio description can be created. One of the bases of this book is an assumption that an audio description script can be also created through translating it from a foreign lan-guage to one's mother language. When investigated from this perspective, audio description can be also considered as **interlingual translation**, and also some-how as **indirect translation**, since the translator who translated a script does not translate the original text, which is the picture, but its intersemiotic translation.

Audio description is also considered to be an **intralingual** translation. This ar-gument is presented by Psiuk, who writes that an adaptation of an audio descrip-tion script – after translating it literally from a foreign language to one's native language – allows us to classify audio description also as intralingual translation:

> Similarly to the case of audiovisual translation, it is necessary to apply various tech-niques in order to convey the sense of the source text, and also to adjust the text to cul-tural and technical conditions (e.g. time limit). The translated text is usually subject to a major compensation in the adapting process which influences its final shape. Therefore it is a case of intralingual translation, because the text somehow does not suit the culture or the conditions in which it is supposed to work. Sometimes the diversity of cultures demands additional explanation, an extension, which is also called overtranslation or

explicitation. No matter which of the techniques is applied, its final effect is intralingual translation (2010: 26, trans. A.M.).

Another sign of intralingual translation can be the fact that during the process of sound production, the written text is changed into the read text, and a translation from written signs to phonic sounds takes place. (Bourne & Jimenez Hurtado 2007: 176).

1.2.3 The consequences of perceiving audio description as a subject of translation studies investigation

Placing audio description in the scope of translation studies has measurable practical consequences. First of them is leaving the so called objectivity criterion, which is very often mentioned in the context of the standards of audio description creation. The topic of objectivity appeared in many discussions during scientific conferences on audiovisual translation and audio description, and it was subject of heated disputes.

The supporters, or the then supporters of the above mentioned criterion claim that audio description should be objective, which means that it should avoid any interpretation and present only the things which are visible on the screen, and that it should not tell the story, but describe the scenes that create it (Jankowska 2008: 230; Snyder 2008: 196; Strzymiński & Szymańska 2010: 11 and Chmiel & Mazur 2011: 292;). They claim that only objective description enables blind viewers to interpret the visual content on their own, and allows them follow the developing thread of a story, and hear and understand what is happening on the screen (Strzymiński & Szymańska 2010: 11). Some people who work with audio description theoretically and practically contradict these views and claim that a hundred per cent objectivity is impossible, and that audio description is always subjective, since it is a choice made by a particular audio describer (Butkiewicz, Künstler, Więckowski & Żórawska 2012: 5). The idea of perceiving audio description as a subject of translation studies is also a part of this movement. According to Sabine Braun:

> The multimodal nature of source and target text in AD also has an impact on the interpretive element, which is present in AD as well as in any other form of translation (…). Just as the production of any translation (i.e. target text) is based on the translator's interpretation of the source text (…), so is the creation of an AD script based on the audiodescriber's interpretation of the audiovisual source (2008: 17).

Approaching audio description as a type of translation also allows us to leave the concept of rules for the concept of strategies, which can vary like in the case of translation (Jankowska & Szarkowska 2011, Mazur 2014). Thinking about audio description as a strategy, connected with the belief that the rules in force are a collection of revealed truths, anecdotic commandments and the so far applied practices (Udo & Fels 2011 and Fryer 2009) rather than a collection of hints based on a thorough research on the needs of the target recipients, opened the door for experiments aiming at the creation of new types of audio description. Thanks to the rejection of the rules of alleged objectivity, we can speak of creative audio description, whose scripts are created on the basis of film scenarios, and use colourful language, metaphors and descriptions of emotions (Jankowska & Szarkowska 2011 and Szarkowska 2013), which includes the elements of film language (Fryer 2009) which are forbidden by the current rules, or from the perspective of one of the protagonists (Udo & Fels 2011). It is worth noticing that all of the "experimental audio descriptions" were created in the frames of research projects whose final stages were reception studies, which proved the enthusiastic reception of the target audience.

The aim of this book is not to criticise or challenge the existing guidelines, but rather to describe the current audio description practice. While the importance of a document laying down guidelines for inexperienced audio describers is undeniable, in order to maintain a thorough and comprehensive viewpoint, the controversies raised by the existing documents must be addressed. Audio description standards established by international researchers (Rai, Greening & Petre 2010) and their Polish counterparts (Strzymiński & Szymańska 2010 and Butkiewicz, Künstler, Więckowski & Żórawska 2012) suggest that, despite the message to the contrary, audio description is not a difficult art. All authors need to do is to follow the rules, familiarise themselves with the work and describe the image on the screen, focusing on the most important plot elements. Further, describers must not interpret the image or add personal conclusions, and remember to fit the description between the dialogue (Strzymiński & Szymańska 2010: 20). From the researchers' point of view, the process is much more complicated, as both the method of creating the description and its content is subjected to critical analysis. This beggs the question whether the currently practised descriptive method, which ignores the film's narrative structure, does not diminish the pleasure derived from its screening (Malzer-Semlinger 2012). Another controversy centres on the proper method of selecting plot-relevant elements, as it is virtually impossible to describe everything that goes on the screen (Remael & Vercauteren 2007 and Vandaelea 2012). This is especially pertinent, as research showed that audio description scripts focus on elements, which did not attract the attention of most viewers without any visual impairments (Mazur & Chmiel 2011).

It is believed that this is due to improper analysis of the film material, which is simply viewed several times by audio describers who lack the ability to read the language of cinema. Consequently, the end product, often referred to as 'objective', is in fact superficial and fails to take account of the narrative structure of the movie or its intent (Orero 2012: 26).

As far as we know, only the publication of the first British *Guidance on Standards for Audio Description*, issued in 2000 by the Independent Television Commission, was preceded by extensive research carried out between April 1992 and December 1995 by the European AUDETEL consortium (Independent Television Comission 2000). The studies involved several phases. During the first one, visually impaired persons were asked to complete a survey on their TV habits and the problems they encounter while consuming TV content of different genres. Next, two hundred visually impaired persons were presented with audio described films and programmes and asked for comments. Finally, between July and November 1994 some one hundred visually impaired persons were asked to watch audio described films for seven to ten hours a week as part of a pilot study. The participants were surveyed about the quality of audio description on a regular basis, with the questions tackling both its artistic value and the technical aspects related to handling the special set-top box enabling access to the additional audio described sound track (Independent Television Comission 2000).

The guidelines or standards existing in many countries, including Poland, are often established with reference to personal experience of their creators or to similar guidelines used on foreign markets, which are often derivative of other sources. Barbara Szymańska and Tomasz Strzymiński, who wrote the original *Polish Audio Description Standards for Audiovisual Productions*, describe this process as follows:

> The work on the standards was divided into several stages. The standards were based on partial studies and extensive experience gathered by the founders of the Audio Description Foundation with respect to audio description creation and implementation, as well as teaching. The document was also inspired by standards applicable in Great Britain, the United States, Canada and Spain (2010: 5, trans. A.M.).

The authors of the *Rules of Audio Description Creation* used a similar strategy:

> The rules presented below, governing audio description creation for blind and partially sighted persons, were based on our editing experience dating back to 2007, repeated consultations with our audiences and typhlopedagogues, as well as analysis of foreign standards (Butkiewicz, Künstler, Więckowski & Żórawska 2012: 2, trans. A.Ch.).

The choice of words used to refer to these documents is also interesting. Polish authors frequently call their publications "rules" or "standards", implying that the document constitutes a set of laws and regulations which must be followed and should not be flouted. The authors themselves agree, for instance by emphasising the following:

> These standards represent a systematic set of professional rules and guidelines for audio describers of audiovisual productions [...] [Their purpose is to] standardise the practices used by professional audio describers, harmonise audio description norms and the rules governing their implementation with rules and norms in place in other countries, and to promote uniform rules of conduct for audio description professionals [...] (Strzymiński & Szymańska 2010: 5, trans. A.Ch.).

On the other hand, the corresponding British document is referred to as "guidance" or "guidelines". By definition, guidelines can be either followed or ignored. The authors of *Guidance on Standards for Audio Description* argue the following:

> Among a large volume of valuable experience, the research revealed that there are many definitions of a successful audio description, not merely because describing styles differ, but because there are many fundamental differences in audience expectation, need and experience. (Independent Television Comission 2000: 4).

This perspective, although apparently forgotten by the authors of 'rules' and 'standards', bears resemblance to the perspective which places audio description creation in the context of translation studies. With this approach, audio describers do not have to follow rules to create the one and only 'proper' audio description, but are free to choose a strategy for developing different kinds of scripts for the visually impaired.

1.3 The history and the current state of audio description

The aim of the next chapter of this book is to present the history of the beginnings and the development of audio description since the birth of this idea to the first moment when it was used in practice as a service for the visually impaired. We will also describe both the historic and the current situation in the United Kingdom and in Poland.

We are aware of the fact that this book is not a historical investigation. Therefore, the historical perspective was limited to the time the notion of audio description started to function. The historical and current perspective of audio description's development in Poland and the United Kingdom is limited to the events connected with audio description in cinema, television and digital display systems. We also investigated such issues as legislation, standardisation and the availability of courses, since they have a great influence on the level of the current development of audio description in both of these countries.

In order to understand the sense of the research that was conducted, it seems necessary to understand the origins of the idea of making visual culture available to visually impaired people. In order to understand why the scripts from the United Kingdom are the ones to be translated into Polish, one should know what are the differences between the development of audio description in these two countries.

1.3.1 The beginnings of audio description

It is commonly accepted to say that audio description exists since people without visual impairment started to describe the world to the visually impaired. Therefore, one can state that it has always existed. However, the formalised audio description – understood as conscious efforts of individuals, organisations or institutions whose aim is to make the products of audiovisual culture available to a broader audience of visually impaired recipients – is a relatively recent phenomenon (Greening & Rolph 2007: 131). It is commonly claimed that professional audio description was invented in the USA, since its theoretical bases were described there in the mid-1970s. However the first documented project that used a narration added to an audiovisual work was realised in Spain in the 1940s already.

According to Pilar Orero (2007b: 112), the first attempts of adding audio description to films happened in Spain in the mid-1940s already, which was even before the Spanish public television TVE broadcasted its first programme. Jorge Andares, one of the first Spanish audio describers, recollects during the interview conducted by Pilar Orero (2007a: 179) that those first attempts were related to theatrical films, transmitted live from a cinema via radio, once or twice a week. According to Arandes, the pioneer of those programmes was Gerardo Esteban, an employee of Radio Barcelona, who trained Arandes in this profession. As a part of the project they presented such films as *Marie Antoinette* (dir. W.S. Van Dyke, 1938), *Gilda* (dir. Charles Vidor, 1946) or *Francis* (dir. Arthur Lubin, 1950). They also attempted to add an additional narration to *Tosca*, an opera transmitted from *Gran Teatre del Liceu* in Barcelona. The project was stopped in the mid-1950s due to a new policy introduced in the public radio (Orero 2007a: 186). Although the achievements of Gerardo Esteban and Jorge Arandes are very impressive, one should notice that their project was not aiming at visually impaired people. Obviously, those people benefited from this kind of programmes. Arandes himself emphasises it and recollects many expressions of gratitude sent by the National Organization of Spanish Blind People (ONCE)[8] and the fact that in the course of time the greeting directed at the listeners started to include a special invitation for the blind (Orero 2007a: 185). Nevertheless, the aim of this programme was to make cinema available to those citizens of Spain, who could not get there. And this is the reason to ask whether the motives which began the above mentioned audition justify calling the first Spanish attempts audio description in the current meaning of the term.

The beginnings of audio description understood as a technique of making audiovisual products available to the visually impaired date back to the 1960s, when Chet Avery, a blind employee of the American Ministry of Education, learnt about a project of creating special subtitles that enabled deaf people to watch television, which was being implemented in one of the departments:

> Avery, who postulated an establishment of a project that would describe films for the visually impaired, tried to convince blind people organisations to join an active campaign in favour of adjusting television to their needs. However, his efforts turned out to be ineffective, because, among other reasons, of the fact that at that time the blind people organisations preferred to concentrate on organising working posts for their members, rather than on audio description (Jankowska 2008: 234, trans. A.M.).

Professor Gregory Frazier, the author of theoretical bases of audio description for the blind, which he called "Eye in the Ear" at that time, is an important, although often neglected, figure in the history of audio description. According to the sources, Gregory Frazier came up with this idea when watching the Western *High Noon* (dir. Fred Zinnemann, 1952) together with his blind friend, who asked him to describe what was happening on the screen. Many years later Professor described that moment as "the most exciting experience in his life", saying that at that moment "the light bulb went on" and he himself "was a changed man" (Thomas 1996). Frazier, inspired by this episode, went back to San Francisco State University to study broadcast journalism. In 1957 he defended his master thesis titled *The Autobiography of Miss Jane Pittman: An All-audio Adaptation of the Teleplay for the Blind and* Visually Handicapped, which was the first work that included a description of theoretical ground for audio description (Frazier 1996). Frazier assumed that audio description should be applied to a greater extent than spontaneous and simultaneous whispering, and it had to be recorded and then played in a cinema through wireless headphones, and via an additional sound canal in television. The concept outlined by Frazier generated interest of the Dean of Creative Arts at San Francisco State University – August Coppola, who

---

8   *Organización Nacional de Ciegos Españoles.*

came from a family of filmmakers. In 1987, Gregory Frazier and August Coppola established *AudioVision Institute* at USF and later also a non-profit organisation taking care of audio description promotion. The première of the first film audio described by Audio Vision Institute took place during the Film Festival in Cannes (Asimov 2009). The institute worked on audio description for films, theatre shows and TV programmes, which were enthusiastically received by blind people (Thomas 1996). The Institute prepared audio description for such titles as *Tucker: The Man and His Dream* (dir. F.F. Coppola, 1988), *Indiana Jones and the Last Crusade* (dir. S. Spielberg, 1989) and *Casablanca* (dir. M. Curtiz, 1942) (Thomas 1996 and Frazier 1996). Frazier asked the Ministry of Education for some financial support for audio description and the introduction of a necessary legislative which would enable the implementation of audio description in the whole country. Although he tried two times, his applications were rejected, and Frazier's activity was limited to the San Francisco region (Frazier 1996).

Margaret and Cody Pfanstiehl, a married couple from Washington, who established the foundation *The Metropolitan Washington Ear*, were active. Initially, the activity of the foundation was focused on running a radio station for the visually impaired, where current magazines and books were read aloud (Washington Ear 2002). It is commonly thought that Pfanstiehls were the first to use audio description for a greater audience during the show *Major Barbara*, which took place at Arena Stage theatre in Washington in 1981 (Jankowska 2008: 235). During this and subsequent screenings audio description was read live and received through sound amplifiers available in the theatre, which were originally intended for hearing impaired people (Jankowska 2008: 235). In 1982 Pfanstiehls started working with the TV station *Public Broadcasting Service*[9] (PBS), to start a simultaneous audio description service that was broadcasted to the show *American Playhouse* (Audio Description Coalition 2009). In 1986 Washington Ear produced the first audio description for museums, recorded on tapes. Part of this project were the recordings describing the Statue of Liberty, Fort Clinton and two national monuments supervised by the federal agency's National Park Service (Audio Description Coalition 2009 and Künstler 2007: 58). Between years 1987 and 1988 foundation *The Metropolitan Washington Ear* started collaboration with PBS, *WGBH Educational Foundation*[10]. The effect of this collaboration were next seasons of *American Playhouse* TV series, where audio description was

---

9   American network of 354 non-profit public stations.
10 Educational Foundation at the pubic broadcasting company WGBH (West Great Blue Hill).

transmitted via satellite television, using SAP (Secondary Audio Programme) stereo transmission, which means an additional channel for the transmission of the second track (Typhloinformation Laboratory of the Institute of Informatics 2002). Thanks to such a solution, the visually impaired could hear an additional description, when the audience without vision impairment watched the film without audio description.

An effect of this collaboration was *Descriptive Video Services* – an internal unit created by *WGBH Educational Foundation* in 1990, which was delivering audio description for television (Jankowska 2008: 235). After two years only this unit was transformed into Motion Picture Access Project (MoPix), which has been active until today. MoPix creates audio description and subtitles for the deaf, as well as technologies which are necessary to receive them in television, cinemas, national parks and Disney amusement parks (MoPix Motion Picture Access).

Another important date for the development of audio description in the USA is year 1989, when blind James Stovall and partially sighted Kathy Harper established a TV network called *Narrative Television Network* (Jankowska 2008: 235). Its founder says that he came up with the idea of adding narrative description to films when he lost vision at the age of 29:

> I was scared to think about going outside my front door. [...] I thought: This is it for me, this is the rest of my life right here, with my little radio and my telephone and my tape recorder; and I'll never leave this room again. [...][I was watching] an old Bogart film entitled *The Big Sleep*. I had seen it enough times that through the sounds alone — and my memory — I was able to follow along fairly well until, in the middle of it, somebody shot a gun, and someone else screamed, and a car sped away, and I forgot the plot of the movie [...] [Then I thought that] somebody needed to provide a service that would make tapes "watchable" by blind people, by adding narration that would explain what was happening onscreen (Farnham 2000: 209).

The idea of *Narrative Television Network* station, which is still active, is unique in the world – its main premise is to broadcast only programmes with audio description that cannot be turned off (Jankowska 2008: 238). Interestingly, it is one of the reasons of the unusual success of the network – most of its audience are people without visual impairment, who benefit from an additional narrative track, since they can do other things simultaneously to watching TV (Jankowska 2008: 238).

Rapid development of the services providing television with audio description happened at the turn of the 1980s and the 1990s – the programmes including an additional narrative track were watched by circa 30 million of households (The Institute of Informatics of the Jagiellonian University 2002). This was probably the reason why in 1990 the American Television Academy appreciated the efforts of American audio description pioneers, and awarded *AudioVision Institute, Metropolitan Washington Ear, Public Broadcasting Service, WGBH Educational Foundation* and *Narrative Television Network* with a team Emmy prize. In spite of that award and the efforts of many organisations, audio description did not appear in any commercial television, which was emphasised by Gregory Frazier in January 1996 in his letter to Federal Communications Commission, which he wrote half a year before his death.

It is difficult to evaluate who should be named the pioneer that "invented" or "discovered" audio description. Probably, it is better to adhere to the already mentioned statement that audio description exists since people without visual impairment describe the world to the visually impaired. However, it is unquestionable that at the beginning of audio description the USA was the unchallenged pioneer of audio description implementation.

1.3.2 The history of audio description in the United Kingdom

The United Kingdom was the first European country that used audio description and the idea was initiated there in the mid-1980s. (Jankowska 2008: 235). In 1983, in the small Robin Hood theatre in Averham, the first European theatre performance with audio description took place (Orero 2007b: 112). Soon, audio description was used in British cinemas – audio described films were regularly projected in Chapter Arts Centre in Cardiff (Jankowska 2008: 236).

The crucial moment in the history of British audio description was the year 1991, when the Independent Television Commission[11] established the AUDETEL project (Audio Described Television), which was a European consortium of the national organs' representatives supervising television broadcasting companies, television broadcasting companies and organisations of visually impaired people (Jankowska 2008: 236). According to Greening and Rolph (2007: 128–129), in the course of the actions taken, a decoder that enabled a broadcast of audio description in analogue television was developed and tested. A work on the project of an act obliging to introduce audio description was also started. However, the project was suspended due to the implementation of an act from 1996, which in fact obliged broadcasters to include audio description in 10% of the programming, but this obligation was supposed to be brought into force not until 10 years after the introduction of the terrestrial digital television (Greening & Rolph 2007: 128–129). In the meantime, many attempts to make television

---

11 *Independent Television Comission,* a British state organ that supervises television broadcasters.

available to the visually impaired were conducted, but technology proved an insurmountable obstacle. Finally, audio description was made available to a broader audience for the first time in 2002, via a digital platform called *British Sky Broadcasting*. This event was even more unusual, since in 2002 the introduction of audio description to television was not a statutory obligation. Greening and Rolph (2007: 128) underline that the implementation of audio description in television was significantly accelerated by two crucial legal acts. In 2003 *Communications Act* was brought into life, which obliged broadcasters to add audio description to the programmes broadcast via digital and satellite television as well. In 2004 Ofcom[12] issued *Code on Television Access Services.*, in which it was recommended that the broadcasters implement audio description five years after the introduction of digital television.

Regarding the cinema, after first attempts in Cardiff, Napier University, *Royal National Institute of Blind People* (RNIB), International Audio Description Agency and the organisation Glasgow Film Theatre started collaboration. An effect of this collaboration was the *Cinetracker* facility, which enabled playing a recorded audio description track simultaneously with a film and releasing the sound to wireless headphones. Unfortunately, the project was abandoned due to some synchronisation problems (Greening & Rolph 2007: 133). The first genuine audio description screening took place in 2002 in a cinema. The screening of the film *Harry Potter and the Sorcerer's Stone* was possible thanks to the DTS technology (*Digital Theatre Systems*), which enabled recording of six multichannel audio tracks not on a film tape, but on a DVD which is synchronised with the film (Greening & Rolph 2007: 133 and Jankowska 2008: 233). Thanks to it, one of the tracks can be left for audio description, which is completely synchronised with a film and is accessible via headphones.

1.3.3 The current state of audio description in the United Kingdom

The United Kingdom is still the leader of audio description implementation. Joan Greening, one of the first British audio describers, emphasised the fact that

---

12 Office of Communications – British state organ established in 2003 that supervises television broadcasters and which took over the rights of the five previously existing institutions: Broadcasting Standards Commission, Independent Television Commission, Office of Telecommunications, Radio Authority and Radiocommunications Agency.

it is possible not only thanks to legal regulations, but mostly thanks to enthusiastic and emphatic attitude of the society, broadcasters and distributors, as well as many social campaigns organised by non-governmental organisations (2010).

The situation of audio description in television seems very good. Currently, audio description is available both in digital television and in the video on demand service. According to the legal regulations, television broadcasters in the United Kingdom are obliged to add audio description to a part of their programming – a per cent set for a given station is proportional to its size, but in none of the cases it exceeds 10% of programming. According to the report of the Office of Communications, in 2012 as much as 96 stations, i.e. circa 90% of all broadcasters, broadcasted audio described programmes (Office of Communications 2012). The data included in the report suggest also that the broadcasters not only meet the requirements dictated by law, but also exceed the set percent threshold – some of them, like BBC stations, or Sky satellite platform stations, audio describe from 20 to 36 percent of their programming. Audio described programmes are also available in BBC and Channel 4 Internet services.

British cinemas are not legally obliged to use audio description. However, according to Joan Greening and Deborah Rolph (2007: 134) – thanks to the enthusiasm of the owners of cinemas, distributors and software producers, the introduction of audio description to cinemas proved successful. As early as 2007 audio described films could have been watched regularly in 300 cinemas, over 80% of the mainstream cinema films were audio described, and such film studios as *Buena Vista International* and *Warner Brothers* added audio description to all of their productions (Greening & Rolph 2007: 134).

The amount of DVDs and Blu-rays available for blind and partially-sighted people is also admirable. Until January 2013 over 500 titles were released on digital storage devices (Royal National Institute of the Blind 2013).

The situation of the availability of audio description trainings in the United Kingdom is also very good. The trainees can participate in professional postgraduate studies which are completely devoted to audio description (University of Surrey and City University of London), or participate in audio description class which is a part of master or postgraduate studies (University of Wales, University of Sheffield, London Metropolitan University, University of Leeds, Roehampton University and Dublin City University). One can also participate in a training organised by independent *Audio Description Association* and its partner which is a network of state professional colleges (Jankowska 2008: 241). Audio describers are also trained by audio description production companies.

The United Kingdom is also one of the first countries that prepared a document that includes rules concerning audio description creation. *ICT Guidance on Standards for Audio Description* was commissioned by the Independent Television Commission (ITC). The first version of the document was issued in May 2000. Currently, a new version is available, which was updated by the Office of Communications in 2006.

1.3.4 The history of audio description in Poland

One could write that audio description appeared in Poland at the beginning of the 21st century and insert a full stop with an almost clear conscience. Actually, the term *audio description* (Pl. audiodeskrypcja) that is eagerly used today appeared in Polish quite recently, namely in 2006. One of the first Polish audio describers, Krzysztof Szubzda, recollects that it was a time when no Polish website appeared in Google after entering the term *audiodeskrypcja* (Szarkowska 2008: 132). However, if one entered the term *typhlofilm* (Pl. *tyflofilm*) the search would be probably much more satisfactory.

The initiative of making films available to blind and partially sighted started to germinate at the end of the 1990s already, which is only a couple of years later than it happened in the United Kingdom – the European pioneer. At that time the currently used term *audiodesckrypcja* did not exist in Polish, and therefore "screening versions enhanced in such a way that they are available to the blind" (Szczepański 2001) were called typhlofilms from the word *typhlos* (Gr. *blind*).

The first public typhlofilm screening for the visually impaired took place in 1999 in Muszyna, during a conference organised by the Central Library of the Polish Association of Blind People[13] (Ciborowski 2010). The audience could watch the film *Ekstradycja III* recorded on a VHS tape (Szczepański 2001). The project was initiated and executed by Andrzej Woch – a blind employee of the Institute of Informatics of the Jagiellonian University, connected with the Cracow Typhlogical Society and the owner of Handisoft agency (Jankowska 2008: 242). Ciborowski describes this event as follows:

> The first public screening of Polish audio description took place in 1999 in Muszyna, during a conference organised by the Central Library of the Polish Association of Blind People. It was a screening of a film recorded on a VHS tape. The method was to record an audio description text on the 1/4th height of the tape. It allowed to play the film and the commentary created for it simultaneously – says Andrzej Woch, the creator of first Polish typhlofilms. This method was used to prepare 20 audio described films. Andrzej

---

13 Centralna Biblioteka Polskiego Związku Niewidomych.

Woch, as the first person in Poland, became interested in the creation of descriptions for the blind in 1998. At that time such productions were called *typhlofilms*. The Polish father of audio description created also a special module software comprising of 3 tools: one to create text, the second one to record audio description and the third one to play the film and audio description at the same time. This software was used to create films for the blind, in a digital version already, as early as the beginning of the 2000s (2010).

After the success of the Muszyna screening, Woch prepared next films recorded on VHS tapes, which were ordered by the Polish Association of the Blind (PZN)[14] and financed by the Ministry of Culture and Art (Szczepański 2001). Then the films were available in the Central Library of the Polish Association of the Blind (Szarkowska 2008: 22). Various sources give different numbers of those films, from several to fifty films. Among the available titles one can mention *Teddy Bear* (dir. S. Bareja, 1981),, *With Fire and Sword* (dir. J. Hoffman, 1999),, *The Deluge* (dir. J. Hoffman, 1974), *Dogs* (dir. W. Pasikowski, 1992), *Sara* (dir. M. Ślesicki, 1997), *Sexmission* (dir. J. Machulski, 1983), *The Daddy* (dir. M. Ślesicki, 1995), *Vabank* (dir. J. Machulski, 1982), and *Rio Bravo* (dir. H. Hawks, 1959), *The Magnificent Seven* (dir. J. Sturges, 1960), *Titanic* (dir. J. Cameron, 1997), *The Scent of Woman* (dir. M. Brest, 1992) or *The Horse Whisperer* (dir. R. Redford, 1998).[15] The last five films deserve particular attention, since it was the first and, until 2011, almost unique initiative of adding audio description to foreign films. Additionally, Handisoft agency run by Woch prepared also twelve films (including *Boys Don't Cry* and *Fast Lane*) on digital storage devices that could be used both on computers and DVD players. Thanks to this project Andrzej Woch was ahead of his time again, since he prepared films read by a voice talent actor and by a speech synthesiser (Link 2002).

The films created by Woch are still called typhlofilms, since the way in which they are created differs significantly from the commonly approved rules of audio description. The main difference concerning the production of typhlofilms was the use of a so called freeze frame, which meant stopping the picture and the sound in order to add description for the blind and partially-sighted people in the places, in which it would not fit in a normal tribe of a film screening. It seems that such a way of film preparation was not criticised at the beginning. The stories told by blind people and teachers from schools for visually impaired children, as well as a few press reports let us conclude that the films prepared by Woch generated real interest (Szczepański 2001). In the course of time, the idea of typhlofilms aroused more and more controversies. One should emphasise that characteristic features of typhlofilms did not allow to use them in TV, and in a cinema they would demand

---

14 Polski Związek Niewidomych.
15 *Miś, Ogniem i Mieczem, Potop, Psy, Sara, Seksmisja, Tato, Vabank.*

an organisation of a separate performance for the blind, and when it comes to digital devices, it would be necessary to add a separate picture and audio track that would aim at visually impaired people only. Some people claim that the fact that it was impossible to use typhlofilms in cinema, television and on digital devices hindered the audio description development in Poland. Authors' unwillingness to interfere in their work to such a great extent was another argument. However, a part of the blind audience claims that thanks to this solution they were given all of the information needed to understand a film, and when they could choose between insufficient information and a freeze frame, they chose the latter.

Eventually, the typhlofilms project was suspended. The possible reasons are both the lack of sufficient interest of blind people, and the insufficient financing of this cost and time consuming enterprise. Another important factor was probably the film distribution problem or the fact that they were not popularised enough in the blind people environment (Jakubowski 2007). In order to borrow a film, people needed to visit the Central Library in Warsaw and fill in a special paper. This means that the films were actually available to a small group of blind people who lived in the capital.

At this point it is worth mentioning that currently in Poland a so called "extended audio description" starts to function. It is used in the cases, where there is no space for the classic audio description that does not interfere with the time line. It is used for promotion films or election spots (Foundation for Audio Description Progress Katarynka 2014). Its use is also recommended in films, in which "the dialogue track is dense" (Zadrożny 2010, trans. A.M.). Taking into consideration the fact that the so called "extended audio description" is based on stopping a film, which is a freeze frame, we can risk saying that maybe the career of typhlofilms has not been ended yet.

The date of reactivation or the second beginning of audio description in Poland is considered to be the 27th of November 2006 (Jankowska 2008: 242). On that day the film *Extras*[16] (dir. M. Kwieciński, 2006) was displayed in the *Pokój* cinema in Białystok. The initiator of that event was Tomasz Strzymiński, the current president of the Audio Description Foundation. The audio description script was prepared by a cabaret artist Krzysztof Szubzda, who also read the script. This screening started a pilot project: from February to December 2007 five audio described films were shown[17] in cinemas in Białystok, Poznań, Elbląg

---

16 *Statyści.*
17 Following films were shown: *Extras* (dir. M. Kwieciński, 2006), *Ryś* (dir. S. Tym, 2007), *U Pana Boga w ogródku* (dir. J. Bromski, 2007) and *Ice Age II* (dir. C. Saldanha, 2006).

and Łódź. All of the films were displayed during specially organised screenings, audio description was read live and heard via a public channel, since no headphones were used (Jankowska 2008: 243).

The first cinema screenings in accordance with world standards took place as a part of the 32nd Gdynia Film Festival (Jankowska 2008: 243)[18]. The audio description track was read live and it was heard only by people who were interested in it and took headphones sets used for simultaneous translation. At this point it is worth mentioning that first cinema screenings with recorded audio description that was played via headphone took place in Kraków, during The International Film Festival Etiuda&Anima in November 2010, thank to the efforts of the UNESCO Chair of the Jagiellonian University[19].

The first screenings with audio described foreign films took place in 2007 in Bydgoszcz[20], thanks to an initiative of the teachers from L. Braille Special Training and Education Centre (Jankowska 2008: 243). The films presented during the show were translated by means of subtitles, therefore it was necessary to read aloud both audio description and the subtitles (Jankowska 2008: 243).

The year 2007 proved ground-breaking also for audio description in television productions. Before a change into terrestrial digital television Polish Television made available part of its materials via an Internet platform. Since the 14th of June 2007 blind and partially sighted people could watch selected Polish TV series[21], available on a special Internet website www.ivp.pl (Jankowska 2008: 243). Episodes were secured with a special code, which one could get for free from the Polish Association of the Blind (Jankowska 2008: 243). Soon, those shows appeared on DVDs and one could buy them. Sooner, however, the first DVD with audio description was released[22] – since February 21, 2008 DVDs with the film *Katyń* directed by Andrzej Wajda were available for purchase

---

18 During the festival the film *The Crown Witness* was displayed (dir. J. Sypniewski, 2007, audio description: D. Jakubaszek) as well as a children's film *Winky's Horse* (dir. M. Kamp, 2005, audio description: A. Jurkowska).

19 During a special show 15 student etudes were displayed (audio description: J. Drożdż-Kubik, A. Jankowska, A. Psiuk).

20 *The Lives of Others* (dir. F. Henckel von Donnersmarck, 2006), *Night at the Museum* (dir. S. Levy, 2006), *Step up 1* (dir. A. Fletcher, 2006), *Empties* (dir. J. Sverak, 2007). Audio description to all of the above was prepared by: Joanna Dłuska and Jacek Knychała.

21 *Ranczo*, (*The Ranch*, dir. W. Adamczyk, 2006-) *Tajemnica Twierdzy Szyfrów* (*The Secret of the Code Fortress*, dir. A. Drabiński, 2007), *Magiczne drzewo* (*The Magic Tree*, dir. A. Maleszka, 2004–2006) *Londyńczycy* (*The Londoners*, dir. G. Zglinski, 2008–2009).

22 Audio description: D. Jakubaszek.

(Jankowska 2008: 243). The first broadcast of an audio described film in a Polish TV station happened in TVP1 on June 13, 2011[23] (TVP 2011).

In July 2011 the Amendment to *Broadcasting Act* was brought into force, according to which Polish broadcasters are obliged to enable the visually and hearing impaired to watch their programmes. The number regulated by the Act is 10% of a quarter time of broadcasting a programme, excluding commercials (The Chancellery of the Sejm 2011:14).

1.3.5 The current state of audio description in Poland

Currently, audio description is available in Poland in a more or less regular way in TV, cinemas and theatres, as well as on DVDs. However, except for television, in which the implementation of audio description is legally regulated, making the product of audiovisual culture available to the visually impaired is possible thanks to individual actions of non-governmental organisations that organise film and theatre shows with audio description and often prepare audio description soundtracks and offer them for free to the cinema and DVDs distributors. The situation, when the broadcasters and distributors try every possible method to limit the spendings connected with audio description implementation and perceive it as a necessary evil, is particularly harmful when one thinks about the already mentioned words of Joan Greening, who emphasised that the success of audio description in the UK was possible thanks to the enthusiasm, sensibility and social awareness of broadcasters and distributors.

As it was described in greater detail in the introduction to this book, the situation in Poland is still not satisfactory, but one should not forget about the progress Poland made in recent years. It is best proved by the words of the author of this book, which come from an article published only six years ago and which are related to the matter of legislation:

> When deciding to implement audio description, Poland has to face a couple of significant problems. [...] At present, there is no official document that regulates the question of implementing audio description in Polish media. However, the efforts are made to include a regulation concerning the implementation of audio description in the non-discrimination act (Jankowska 2008: 244, trans. A.M.).

The Amendment to the Act is undoubtedly a ground-breaking event, although it arouses much controversy. At the moment of finishing this book the work of the Group for the monitoring of the way of implementing the regulations of the

---

23 *The Secret of the Code Fortress.*

Act in relation to handicapped people was in progress. Particular controversy was aroused by the percent figure of the programmes with facilities. During the debate over the problem, the representatives of broadcasters proposed to interpret the act in such a way that the aids for visually and hearing impaired together must constitute at least 10 % of the quarterly time of the programming broadcast, and the proportions between them depend on the broadcaster. Such an interpretation allows them to add to major part of this percent figure the facilities for the hearing impaired, since they are cheaper[24] than audio description. The organisations of the visually and hearing impaired reacted, as well as the Commissioner for Human Rights, and postulated that such an interpretation of the regulations would lead to a threat that those people can be excluded and that the aids mentioned in the amendment should be treated separately. According to such an interpretation, broadcasters are obliged to provide 10 % of programmes with audio description, 10 % of programmes with subtitles for the deaf and 10 % of programmes translated into sign language.

The situation of the availability of trainings has also changed. Currently, the trainings of audio description creation for audiovisual productions, including theatre shows, exhibitions or sports events, are taught by practically all of the foundations that deal with audio description promotion in Poland. Audio description is also present at universities, where it is available at bachelor and master studies in form of an optional class, as well as in form of professional courses or postgraduate studies. At the moment of collecting the material for this book, a class of audio description as a part of audiovisual translation class was offered in the Institute of Applied Linguistics of University of Warsaw, in UNESCO Chair for Translation Studies and Intercultural Communication of the Jagiellonian University and in the Institute of English Studies of the Catholic University of Lublin. UNESCO Chair of the JU offers also professional weekend courses of audio description for films, theatre and museum. Audio description modules are also offered as a part of the postgraduate studies in audiovisual translation at the Department of English Studies of Adam Mickiewicz University in Poznań and the University of Social Sciences and Humanities in Warsaw. Until today the only university that offers studies which are fully devoted to audio description is the University of Białystok, which announced the recruitment for undergraduate studies in audio description and rhetoric. Unfortunately, the head of the studies said in a telephone call that the studies were not open due to an insufficient number of candidates. However, one should notice

---

24 Subtitles for the deaf and hard of hearing, Polish Sign Language (Polski Język Migowy-PJM) or Signed Polish (System Językowo-Migowy).

that the trainings which are offered seem to be too short and superficial to provide a trainee with all of the skills which are necessary for an audio describer, and the graduates do not have many possibilities to gain practical experience in this job.

Another change is the publication of documents on the rules of audio description creation. While in 2008 there was no code of audio description standards (Jankowska 2008: 244), currently there are as much as two: *The Standards of Audio Description Creation for Audiovisual Productions* written in 2010 by the Foundation Audio Description to the order of University of Warsaw and *Audio Description – the Rules of its Creation* written in 2012 by the Foundation Culture without Barriers. However, as it was already mentioned in the introduction, the issue of the usefulness of rules and standards has been recently questioned.

2. Audio Description In Practice

This part of the book will focus on practical aspects of audio description, including the **process of its development** and **strategies** used to create an audio description script. We will also discuss the rules governing **audio description creation** and, finally, explain the **rationale behind audio description**.

2.1 Audio description creation

Many people wrongfully assume that audio descriptions are monosemiotic written texts, forgetting that for their target recipients they represent a polysemiotic text – namely an additional audio track. Considered from this perspective, the script is the result of a multi-staged process involving several parties. According to the available publications (Benecke 2004; Jankowska 2008; Strzymiński & Szymańska 2010 and Butkiewicz, Künstler, Więckowski & Żórawska 2012) the audio description development process is divided into four stages – selection of the audiovisual material, selection of the sound production method, script creation, recording and editing. Subsequent stages of audio description are described below, with particular focus placed on script creation, as this phase is directly linked to the purpose of this book.

2.1.1 Audiovisual material selection

Various studies suggest that docudramas, documentaries, series, newscasts and feature films are programmes which are the most popular with the visually impaired (Evans & Pearson in Jankowska 2008: 228). In theory, audio description could be applied to each of these genres, but practice shows that some of them are not suitable for the purpose (Ofcom 2000; Benecke 2004; and Strzymiński & Szymańska 2010). According to Szymańska and Strzymiński:

Audio description in programmes dominated by spoken word may force blind audience members to constantly select information in order to understand any of it, which is very tiring (2010: 13, translation A.M.).

The same is true for feature films. The ICT guidelines (2000) recommend that audio description should be avoided in case of films with more action scenes than dialogue scenes or with dialogue-intensive films that offer little space between individual conversations. In both cases, audio description obstacles reception instead of facilitating it, as it requires constant attention of the recipient.

Given the above, each film's suitability for audio description should be assessed before the work on the script begins.

2.1.2 Selection of the sound production method

When setting about to create an audio description script, it is important to consider sound production, since the selected method will affect the final script. Obviously, the script may be re-edited following completion, but in the author's opinion writing the script with a specific method in mind greatly accelerates the entire process.

Audio description, understood as a narrative track for blind consumers of visual content, may be delivered in several different ways. As illustrated in the diagram below, audio description scripts can be read either by a voice talent or a speech synthesiser. In both cases, the audio description may be performed live or from a recording.

Diagram 1: Sound production methods used audio description.

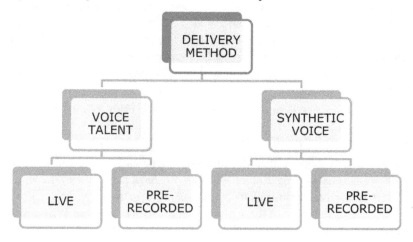

Live audio description[25] is typically used for scene performances, such as theatrical productions, opera shows or other live events, which are unpredictable and require the ability to produce spontaneous responses to live developments. Live reading in cinemas is less common. While at present there is little information about this method being used outside of Poland, domestically live reading during special screenings is the most popular method of audio description delivery. This is because audio description is introduced to Polish cinemas almost solely through the effort of NGOs. In most cases, these organisations lack funding and skills required to provide a professional recording and editing. Consequently, they often decide to hire a voice talent artist to read the script in front of a live audience during the special screening. This solution, however, has more disadvantages than advantages, and is only seemingly cheaper. For instance, let us consider quality. Live reading always involves a risk that the artist may mispronounce a word or miss their cue, resulting in an overlap between the script and the recorded dialogue. Further, live performance does not involve voice mastering, which includes noise removal, voice tone adjustment and finding appropriate volume levels for a given film's sound track. To put it in simple terms – the voice of the live reader is not as nice and silky as the voice of the recording and it is very difficult to strike a balance between the volume of the sound track and the text. Last but not least, there is the matter of finances. The price for live reading of a single movie act ranges from 25 to 35 PLN, VAT-inclusive, whereas a recording costs 80–100 PLN per act[26]. A single performance is therefore cheaper than a recording, but each subsequent screening multiplies the costs, as each time the voice talent has to be hired and remunerated. On the other hand, the same recording, once made and paid for, can be reused at different times and places.

The alternative method of sound production involves feeding the audio description script to a speech synthesiser, which uses complex algorithms to convert written text into speech (Szarkowska 2011: 144). Speech synthesisers can be used for live playback, as suggested by Szarkowska:

> In order to play a film with a synthesiser-read audio description, the user must have speech synthesis software and a media player supporting subtitle reading. The subtitles are loaded into the media player and read by the synthesiser in between the dialogue, according to the assigned time codes. The volume and pace of the read text is adjusted through the media player (2011: 145, trans. A.M.).

---

25 The term refers to reading of a prepared script by an actor in front of a live audience. It should not be confused with live audio description, in which case the script is improvised. Live audio description is used for instance during sports events.

26 Source The Seventh Sense Foundation.

Szarkowska's solution was aimed at increasing the availability of audio described films and was designed for home use by blind audiences:

> From the perspective of the users, i.e. blind or partially sighted persons, synthesiser-read audio description is a solution that requires no additional cost, as most AD audiences already use a speech synthesiser at home or at work. Text-to-speech software available on the market ensures good quality of playback, which means that watching an audio described movie with a computer-read script can be, in fact, entertaining. As opposed to audio described films available online, for instance at the public TV's official website, synthesiser-ready AD does not require a fast Internet connection, as in this case the script is a simple .txt or .sub file which is fed to the software and read by the synthetic voice selected by the user (AVTLab 2012).

Live synthesiser reading has similar disadvantages as live performance by a voice talent. Obviously, the software will not make pronunciation mistakes or miss its cues, but technology too is fallible, which means that, for instance, the synthesiser may crash. To avoid this risk, the synthesiser-read audio description script may be saved as an audio file, using a functionality offered by some of the speech synthesisers. The audio file should be then merged with the film using non-linear editing software, which enables volume and reading pace adjustment. The combined material saved as a video file can be used on any device (for instance a DVD player or a PC) and does not require text-to-speech software, as the audio description has been permanently integrated into the movie's sound track. To the best of our knowledge, this solution has been first used to prepare audio described films for a pilot study carried out for the purposes of this book. Chapter six of the second part of this book includes a section on film preparation, which provides a detailed description of the process of creating a synthesiser-read script, as well as the target video file.

The solution discussed above, involving the use of synthesiser-read audio description, requires rudimentary film editing skills and specialist software, which is why it will probably find application in contexts other than those indicated by Szarkowska, for instance in TV or film screenings. To note, Szarkowska's solution (2011: 144) was and continues to be highly controversial, as many challenge the quality of synthetic voices. However, studies conducted among the visually impaired suggest that they are willing to adopt this solution, both as a temporary and a permanent measure (Szarkowska & Jankowska 2012: 87). This innovative solution, which is attractive both in terms of price and quality, is currently used by the Swiss public television and Polsat TV in Poland.

As mentioned before, the sound production method determines the final form of the script, which has to be adjusted to reflect either the needs of a voice talent, or the requirements of text-to-speech software. When preparing a script which is to be read by a voice talent artist during a live performance or a recording, it is important to maintain a clear layout and to include directions for the reader (Butkiewicz, Künstler, Więckowski & Żórawska 2012). Those should indicate both the times at which the artist has to start reading and the proper manner of reading (for instance accelerated or slow pace of speech). In case of foreign movies, proper names should be written down phonetically. Scripts for live performances and studio recordings differ in the way in which cues are marked. It is recommended that scripts for live audio description include dialogue directly preceding the reader's text – the dialogue should be marked in bold (Seventh Sense Foundation, 2012). The box below presents a sample script for live performances. The dialogue preceding audio description is marked in bold. Information about the reading speed included in the brackets (i.e. [od razu po tytule i dość szybko] = right after the title and rather fast).

Example 1: Audio description script for live reading.

Kosmos. Wirują gwiazdy. Wyłania się galaktyka. Na tym tle pojawia się zielony napis relativiti Media

królewna śnieżka

[od razu po tytule i dość szybko] Zamkowa komnata. Pośrodku stolik. Na nim ogromne kryształowe jajo. Podchodzi rudowłosa królowa w żółtej sukni. Kręci wystającą z boku stolika korbką. Ten obraca się. Wewnątrz jaja pojawiają się kwiaty czerwonej róży. Królowa wpatruje się w nie uważnie. Zaczyna mówić. Na ekranie animacje ilustrujące jej opowieść.

To ciekawy wątek, ale wrócę do niego później.

[szybko] Królowa zakłada sobie i królowi po łańcuszku z wisiorem w kształcie półksiężyca. Król dosiada konia i odjeżdża.

On the other hand, if a script is intended for a recording, the reader's cues should be marked with time codes (see Example 2). The script may, but does not have to, include dialogue directly preceding the reader's lines and should include information about reading speed (i.e. [szybko] = fast).

Example 2: Audio description script for recordings.

Śmierdziel.

00:01:38   [szybko] Wasja bije czapką Pietję. Chłopczyk łapie się za głowę.

00:01:44   [szybko] Wasja wytrzepuje czapkę, kładzie na niej głowę.

00:01:51   Tuż obok leżącego Pietji ląduje pet. Chłopiec łapie go i wkłada go do ust.

Nie dam.

Daj.

00:02:01   Pietja zaciąga się papierosem i wydmuchuje dym na Wasję.
Wasja spogląda na brata przez deski ławki. Pietja mocno zaciąga się petem,
wydmuchując dym na brata. Przewraca się na bok i delektuje papierosem.

The difference in script preparation, evidenced by Examples 1 and 2, is a consequence of the conditions in which the script is to be read. When working in a cinema, the voice talent cannot follow the time code, which is why they need excerpts
from dialogues as cues. In a studio, the same reader can follow the time codes displayed in large letters on the screen. A time-coded script is convenient both for the
reader and the editor who synchronizes the audio description track with the movie.

Audio description editing with a view to using it in text-to-speech software is
slightly more complicated and requires knowledge of specialist software. Szarkowska
(2011: 144) suggests that audio description script should take the form of subtitles
inserted between dialogues using software for subtitle creation and positioning. In
practice, this means that either the completed script is fed to a subtitle editor, divided
into parts and assigned timecodes marking the reader's cues (Szarkowska 2011:144)
or that the entire process of audio description creation takes place in a subtitle editor.

2.1.3 Audio description script creation

Research available to date suggests that there are three audio description
development strategies: individual work (Orero 2012), team work (Benecke 2004
and Gerzymisch-Arbogast 2007) and translation of existing scripts (López Vera
2006; Georgakopoulou 2009 and Jankowska 2010).

The closer we look, the more apparent it becomes that the strategies of individual authors draw on the same pool of basic components which, not unlike
LEGO blocks, can be put together and rearranged at will to obtain a final product, which is a production-ready audio description script. Therefore, it seems
logical that before specific strategies are discussed, we take a closer look at the
phases of the script creation process.

2.1.3.1 Phases of the audio description creation process

Published research, audio description guidelines (Independent Television Comission 2000; Office of Communications 2006; Rai, Greening and Petre 2010; Strzymiński and Szymańska 2010 and Butkiewicz, Künstler, Więckowski and Żórawska 2012) and feedback from Polish audio describers[27] suggest that audio description creation process can be divided into the following phases:

- **Familiarisation with the sound track:** authors are recommended to listen to the sound track several times without watching the movie itself to better understand the needs of the audience and to assess the film's difficulty.
- **Familiarisation with the work:** before commencing work, authors should familiarise themselves with the entire film, its plot and the relations between the main characters and, if necessary, acquaint themselves with specialist terminology.
- **Script creation:** visual information is selected in line with the existing audio description guidelines and specific items are singled out for description. In a nutshell, the script should describe things which appear on the screen and are important for the plot, avoid excessive interpretations and provide answers to basic questions such as "who?", "what?", "how?", "where?" and "when?". The script should therefore contain a description of the characters, plot developments and the way they are presented, as well as the time and place of action.
- **Script synchronisation:** the script should be read aloud to check whether it fits in between the dialogue.
- **Proofreading:** the final version of the script is prepared by the first audio describer.
- **Second proofreading:** the script should be proofed and assessed by another experienced audio describer – in certain cases, it might be also read by the director or the producer.
- **Consultations:** each script should be consulted with blind and partially sighted persons.
- **Final adjustments:** audio description is adjusted according to the suggestions made during consultations and the final version of the script is prepared for sound production.

2.1.3.2 Audio description creation strategies

Audio description creation strategies can be divided into several categories, for instance with respect to whether the process is completed by a team (Benecke

---

27 Based on conversations and e-mails with Urszula Butkiewicz, Izabela Künstler, Irena Michalewicz, Agata Psiuk, Agnieszka Walczak and Przemysław Zdrok.

2004 and Gerzymisch-Arbogast 2007) or by an individual describer (Orero 2012), which is the most common case, or whether the script is created from scratch or translated. The latter strategy is somewhat controversial and has only been used commercialy in Greece. (López Vera 2006 and Jankowska 2010).

Bernd Benecke (2004) was the first researcher to focus on audio describer teams. According to his findings, work in such groups is split into several stages. First, the script is developed by a team of three audio describers, one of whom is visually impaired. Next, the script is synchronised with the natural pauses by being read aloud, and then adjusted by one of the team members. After the final editing is done by a visually impaired person, one of the audio describers prepares the end product. In terms of time required, available data suggest that each team needs five to seven days to write a script for a 90-minute feature film (Gerzymisch-Arbogast 2007).

Individual work is the most common strategy used in audio description creation. Diagram 2 illustrates all phases of the process. Obviously, not all steps apply in all cases, as their final selection depends on the approach favoured by the audio describer. Some professionals who responded to our query confirmed that they always consulted their scripts with blind persons, while others admitted they only did so when working on particularly difficult scenes. The same was true with respect to consultations with other audio describers. On the other hand, all interviewed professionals stressed the importance of studying the described material in detail and recommended that the film be watched at least twice.

Diagram 2: Strategies used by audio describers working individually.

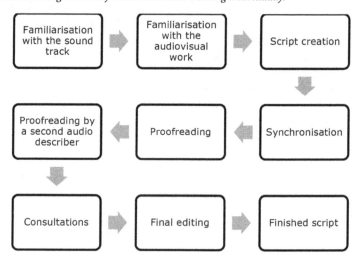

As regards the time required to create a script when working individually, the data are inconsistent. Audio description standards suggest that it takes about a week to write a script for a two-hour film (Strzymiński & Szymańska 2010: 12). Interviewed audio describers needed on average approximately 52 hours to create a script for a hundred-minute feature, but times quoted by individual respondents ranged from 25 to 80 hours, with many describers stressing the fact that the time required to write a script varies depending on the nature of the film in question.

When compared, the two strategies discussed above seem quite similar. In this context, translating a script from a different language appears radically different. As illustrated by Diagram 3, strategies which rely on translation make some of the script creation phases partially or completely redundant. If the source is a finished, thought out script, describers do not have to familiarise themselves with the sound track and the visual aspect of the movie. Given that each described film needs to be listened to and watched at least twice, this represents substantial time savings. Furthermore, assuming that the source text was prepared properly and in cooperation with a blind person, there is no need to consult the target text any further. With a skilled translator, even asking the opinion of a second audio describer or an editor may be unnecessary.

Diagram 3: Strategies used in intralingual script creation.

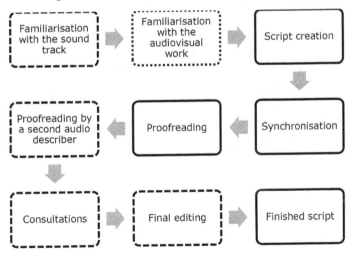

Therefore, it seems that translation allows the audio describer to create a script in four simple steps, which include studying the audiovisual work, translation, synchronisation and proofreading. All of these stages may be performed separately or – as with the voice-over – all at once, with the author translating and synchronising the text at the same time.

Available publications tackling script translation lack any data on time requirements or provide too little information to draw any valid conclusions. Therefore, one of the objectives of this book is to study this matter in detail. The results of an experiment designed to measure the time required to create an audio description script by translating an English text into Polish are presented and discussed in detail in the second chapter of the second part of this book.

2.1.4 Audio describer skills vs. translator skills

As audio description developed into a legally guaranteed service for blind persons, the markets saw a rising demand for audio description professionals. In most European countries, Poland included, audio describers are mostly social activists working with blind persons or audio description enthusiasts. More often than not, they are self-taught and work other jobs, treating audio description as an additional occupation. On the other hand, seeing how the demand for audio description rises, it is possible that a new profession is about to be created.

In Europe, audio description is commonly seen as part of translation studies (see 1.2.). Consequently, prospective audio describers are often trained by academic centres specialising in translation, in particular audiovisual translation. Anna Matamala and Pilar Orero believe that, in fact, translators make the best material for audio describers and suggest that modules on audio description be introduced into syllabuses for prospective audiovisual translators (2008: 329). The same is true in Poland, where audio desctiption is taught mostly by units focusing on translator training or during courses organised by NGOs (see 1.3.5.).

According to the available sources (Navarete 1997, Orero 2005, Matamala 2006 and Díaz Cintas 2007), the most important skills for an audio describer are the following:

- the ability to produce intra- and interlingual intersemiotic translations, as the text may require;
- the ability to sumarise information objectively and correctly and a good command of synonyms and paraphrasing, necessary to ensure that the description meets the time limits while retaining the meaning of the original;
- good command of the native tongue and familiarity with its properties;
- the ability to adjust register to the needs of the target audience and the product;

- the ability to select key information;
- foreign language fluency if working on foreign films with voice over or subtitles;
- familiarity with the theoretic aspects of audio description;
- computer literacy;
- familiarity with audio description software;
- resistance to stress;
- the ability to work against a deadline.

Upon closer examination, the skills above are very similar to those on Jankowska's list of skills required from audiovisual translators, which include:

- translation competence, understood as the product of language skills (fluency, vocabulary range and knowledge of grammar rules), psychological skills (the ability to understand and interpret texts), cultural competence (familiarity with the cultural context of the source and the target) and pragmatic competence (the ability to read the conditions in which the source and the target are performed);
- theoretical competence, defined as familiarity with translation techniques and strategies and the ability to select content of the message;
- technical competence, or the ability to adjust the text to the requirements of a given type of audiovisual translation, including the ability to edit or shorten the text depending on the time and space available;
- technological competence, also known as computer literacy or the ability to operate specialist software;
- soft skills, including resistance to stress or the ability to work against a deadline (2012: 245–251).

Audiovisual translators already have the skills which will make them good audio describers once they complete their training, so it seems logical that they should be more than capable of translating existing foreign-language scripts into their mother tongues.

However, the idea of translating scripts from one language to another gave rise to controversy concerning in particular the translators' skills. When discussing the possibility of translating audio description scripts, Veronika Hyks, a pioneer of British audio description, stressed that even though the skills required from audio describers and translators are somewhat similar, they are not the same (2005: 6). In her opinion, the translator's duty is to provide a faithful representation of the source text in the target language, whereas audio describers need to exercise their judgement to select key visual information and convey it verbally (Hyks 2005: 6). While the notion that a translator does not have to prioritise individual elements of the text is hardly acceptable, the idea of translating audio descripton actually relies on the fact that translator is not required to make decisions as to the content and the method of description, which accelerates the audio description creation process. Hyks also argued that languages differ significantly, as they represent their respective cultures (2005: 6). Similar points were raised in Poland, citing the differences between individual languages, their precision, syntax or even word length[28]. This approach, which questions the very notion of translation, may result from insufficient knowledge of translation theory and practice. Another argument against translation addressed possible inconsistencies between audio description styles and conventions used in the source and target languages (Benecke 2007b, Bourne & Jimenez Hurtado 2007 and Strzymiński 2011). However, this arguments seem to be rooted in personal convictions, rather than empirical evicence since available research proves othewise (see 3).

2.1.5 Recording and editing

As mentioned in section 2.1.2 above, a script may be either performed live or recorded by a voice talent or TTS. As the analysis of reading methods and quality could provide enough material to fill a second book, these matters will not be addressed here, even though we would like to stress their importance. The focus of our analysis will be on audio description delivery, which is crucial given the social aspect of audio description. Diagram 4 presents how audio description performed live or from a recording can be relayed using separate or open channels.

Diagram 4: Methods used to deliver audio description of audiovisual productions.

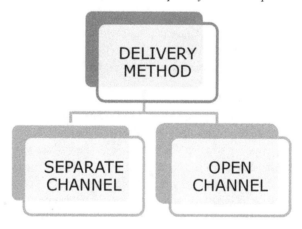

---

28 T. Strzymiński in personal communication with the author.

In other words, audio description may be either delivered via headphones and used only by persons who need it, or played together with the sound track, for instance through the cinema's sound system. In the latter case, all audience members hear the audio description.

The separate-channel method is recommended for public screenings (Jankowska 2008: 232). In this case, audio description is provided via headphones to the interested audience members. This enables audio description to fulfil its role as a social integrator, as persons with and without visual impairments can watch the film together, in a manner that caters for the needs of both groups.

Jankowska (2008: 232) emphasises that audio description was created not only to make audiovisual culture accessible to the visually impaired, but also to allow them to participate in social activities. Consequently, open-channel audio description is not recommended. This solution is cheaper, because it does not require the organiser to rent or purchase the headsets, but does not take into account the social inclusion aspect. There is little chance that an audio described screening will be attended by persons without disabilities. Consequently, visually impaired persons are forced to watch films in their own company, which is not conducive to social integration.

2.2 Audio description beneficiaries

Blind and partially sighted persons are the main beneficiaries of audio description. It is estimated that in modern societies, as much as 20% of the population have eyesight problems (Kalbarczyk 2004: 18). According to the European Blind Union (EBU), which brings together national societies for blind persons from 45 countries, the average European figure may as well be 30% (Jankowska 2008: 229), with 30 million fully blind persons living in Europe alone. The number of persons who experience sight-deficit problems with crossing the street, reading books or watching television is lower than the 20–30% referred to above (Kalbarczyk 2004: 18), but it is likely to increase given the aging society. According to the estimates of British scholars, more and more persons will be affected by sight deficit, as the number of persons over 85 years of age is to double by 2020 (Evans & Pearson 2009: 10).

According to EBU, there are approximately 7.5 million blind and visually impaired persons in Europe (Jankowska 2008: 229). This number, although visibly lower, is still substantial – 7.5 million is the entire population of Lithuania, Latvia and Estonia brought together. Moreover, the figure does not account for everyone, as the EBU's data are approximate and represent the sum of data gathered by national organisations, which only have information on their own members and lack nationwide statistical data. The Polish example shows how big the discrepancy between the two figures may be. According to statistical data, the Polish Association of the Blind has some 70 thousand members, of which approximately 4.5 thousand are fully blind (Strzymiński & Szymańska 2010: 10). At the same time, according to the survey carried out by the Polish Central Statistics Office (GUS) in 2004, there are approximately 146 thousand blind and partially sighted persons in Poland. Furthermore, the European Blind Union defines the meaning of 'blindness' and 'partial sight' very clearly. According to those definitions, a person is blind if they can only read the top letter of the optician's eye chart from three metres or less, whereas a person is partially sighted if they can only read the top letter of the chart from six metres or less. Therefore, we may assume that the number of people whose sight makes it difficult or impossible to cross the street, read book or – more importantly for our subject matter – watch TV is substantially higher. This assumption is confirmed by the GUS survey discussed above, which proves that nearly half a million Poles cannot read the newspaper even with glasses on (Strzymiński & Szymańska 2010: 11).

While the phrase *blind viewer* may sound like an oxymoron, visually impaired persons are active consumers of audiovisual culture, including TV and cinema. This is clearly evidenced by viewership figures among blind and partially sighted people, which were collected in 2008 for the Royal National Institute for the Blind (RNIB) by Roberta Pearson and Elizabeth Evans from the University of Nottingham. The purpose of their study was to address a number of questions concerning TV viewership among the visually impaired and the role TV plays in their everyday life. The study consisted of two stages. The first involved an Internet survey which covered 172 respondents, whereas the second was an in-depth interview with 13 participants carried out in focus groups (Evans & Pearson 2009: 375). The number of participants in the focus group stage was determined by the RNIB – the respondents were people who were available and have previously cooperated with other researchers (Evans & Pearson 2009: 375).

The data collected by Evans and Pearson suggest that blind persons attach great importance to the availability of TV. The respondents consider it to be a source of entertainment (91%). TV also helps them relax (82%), learn about current affairs (77%) and the world (54%), and may be seen as a window on the world (54%) (Pearson & Evans 2008 in Jankowska 2008: 228). Among the visually impaired, the most popular genres are documentaries and feature films (56%), series and soaps (49%), news programmes (35%), feature films (24%), game shows (12%), comedies (8%) and sports broadcasts (6%) (Pearson and Evans 2008, in Jankowska 2008: 228). The researchers also asked about the amount of time spent watching TV. According to the results, 20% of respondents spend less than an hour a day in front of the TV, 43% watch it for one to two hours, 30% – for three to four hours, and 7% – for more than six hours (Pearson & Evans 2008 in Jankowska 2008: 228). Pearson and Evans agree that these preferences are consistent with those expressed by British audiences without impairments.

But the most noteworthy conclusion of the report by Pearson and Evans is the fact that blind persons see television as an important aspect of social life. Firstly, because over half of the respondents (59.6%) watch television with family or friends and, secondly, because – as declared by one of the respondents –television is a part of the mainstream culture and a frequent topic of conversations, so its consumption facilitates integration (Evans & Pearson 2009: 377). The research also suggests than nearly 82% of all respondents talk about programmes they watch, and 66% discuss future plot developments in series they follow (Evans & Pearson 2009: 378). These results led the researchers to the conclusion that for visually impaired persons, TV is a social and cultural space which provides access to information about the world and entertainment and at the same time allows them to build interpersonal relations (Evans & Pearson 2009: 378).

Such a comprehensive study on participation of blind and partially sighted persons in audiovisual culture is yet to be carried out in Poland. Nevertheless, there is no doubt that blind persons in Poland watch television and go to the cinema, even when audio description is unavailable. More than ten years ago, Szczepański noted that:

> Blind people are interested in cinema. They have their favourite movies which impressed them to some extent. They are knowledgeable about acting, directing, scripts, photography and music, and eager to discuss it. They remember the titles of films they saw with their own eyes or after they had lost sight, with the help of their friends. They have their rankings and favourite sets they would like to witness once again. At home, they collect video cassettes and DVDs. They collect CDs with film music and sound tracks and those who use new-generation computers also boast a large collection of MPG files. They can be wonderful storytellers well-versed in anecdotes, behind-the-scenes gossip and other biographical trivia of movie stars. But when they want to see something new, they must always rely on other persons or resign themselves to experiencing their passion only partially (2001, trans. A.M.).

The same conclusions were drawn from numerous private conversations with blind and partially sighted people, as well as the pilot study conducted on January 14th 2010 in the Centre for Children with Special Educational Needs in Kraków, which covered seventeen participants. The study and its participants will be presented and discussed in detail in the second part of this book. However, the results of the study leave no doubt that the blind and partially sighted watch TV and audiovisual productions on digital carriers. On average, the respondents declared that they spend some 2–3 hours a day watching TV, with blind respondents ranking in the lower, and partially sighted – in the upper end of the range. This results should be compared with average viewership figures in Poland. As according to the survey by TNS OBOP, an average Pole watches television for some 3–4 hours daily (TNS OBOP 2011), the difference between the habits of visually impaired and unimpaired audiences seems minor, even more so considering the fact that audio description is not commonly provided by Polish TV stations.

2.3 The importance of audio description

When a visually unimpaired person first learns about audio description, their reaction is either enthusiastic or sceptical, with representatives of the latter group often asking the simple yet pertinent question, namely 'But what for?'. This almost existential question follows from two convictions. Firstly, some of the sceptics believe that the senses of blind and partially sighted people are sharper and therefore they can understand films even without resorting to audio description. Showing this group the importance of audio description is relatively simple – all it takes is to ask them to 'watch' a movie with their eyes closed. Those people quickly learn that hearing does not suffice to understand the plot of a film.

The second group of sceptics follow a different logic. Their 'what for?' is based on a conviction that audio description serves no purpose, as a description can never substitute the image. Their doubts are justified, all the more so because audio description research and implementation receives increasing amounts of public money. The funding forces both theorists and practitioners to carefully consider the *raison d'être* of audio description.

To answer this question, several arguments need to be addressed. Firstly, audio description was not developed by sighted persons. Even though the history of audio description was discussed in subchapter 1.3, it is important to stress once again that both the emergence of the practice and subsequent initiatives related to its implementation were mostly a grass root effort of the blind community.

Secondly, audio description has a long history. Even though audio description as a method of making culture available to the visually impaired developed and became professionalised only in the last ten-twenty years, it is commonly agreed that it has existed ever since someone first described the world to a blind person. But we do not have to go that far. In recent years, the author has had many conversations with blind persons and teachers from special schools, all of which suggest that visually impaired persons have always gone to the theatre or the movies or watched TV. They had been doing it even before audio description was available. At first, their family and friends, guardians and teachers acted as audio describers. As emphasised by Teresa Majeran, one of the teachers from the Lower Silesian Special Educational Centre for Blind and Partially Sighted Children in Wrocław, 'our audio description was maybe not as professional as it should have been, but it was – and still is – there'[29]. Andrzej Woch agrees with this opinion, pointing out that audio description provides more freedom than ever, as now no one has to rely on a neighbour or a son to describe movies for them, or be worried about being reprimanded for talking in the cinema.[30]

Thirdly, respondents in the study by Pearson and Evans declared that they watch TV to feel normal and to participate in the cultural and social life of the nation the same way as others do (2009: 377–379). The last argument is especially important in the light of the Convention on the Rights of Persons with Disabilities (United Nations, 2006) and the EU Audiovisual Directive (European Parliament and Council, 2007), which states that the right of senior citizens and persons with disabilities to participate in social life is inseparably linked with provision of audiovisual media services.

Last but not least, when all rational arguments fail, we can always refer to enthusiastic and lively reactions of the target audiences. Audio description cannot be challenged if blind persons think it is 'the best thing since bread came sliced' (Pearson & Evans 2008 in Jankowska 2008: 245) and after seeing an audio described film admit that they are 'very moved, because for the first time in 11 years they could watch a movie by themselves, and now cannot stop thinking about it and discussing it with their friends'[31].

---

29 In personal communication with the author.
30 In personal communication with the author.
31 Wojciech Figiel in personal communication with the author.

II. Part Two

The second part of this book presents the results of the research comprising of three research experiments, which were undertaken in order to verify whether the strategy of translating audio description scripts is cost-effective, when it comes both to the possibility of saving time, and at the same time decreasing its cost, and fulfilling the expectations of the visually impaired viewers. At the beginning the **state of knowledge** concerning the possibilities of applying the strategies of translating audio description scripts and the **research methodology** will be presented. Next, in three subchapters, the research experiments will be presented: the first experiment – a **time-consumption analysis,** the second experiment – **a pilot study,** and the third experiment – a **cognitive comparative analysis of the scripts.** The parts describing subsequent experiments are comprised of modules describing the research procedure, as well as presenting and summarizing the results of the experiment. At the end of each part we present partial conclusions concerning a given experiment. General conclusions are presented at the end of part two.

3. State of the art

At the time of collecting materials and writing this book, the possibility of applying the strategy of translating audio description scripts has not generated interest of many researchers. This topic was investigated in only two master dissertations and four articles, only half of which are scientific articles, based on results of a scientific research.

The first publication in which the possibility of translating audio description scripts was investigated was written in 2005. The author of a short article titled *Audio Description and Translation. Two related but different skills* is an audio describer with many years' experience – Veronika Hyks. The author argues that, although translation of scripts as a means to save time and increase the accessibility of audio description is a tempting perspective, the process of translating and adopting an already written script can take more time than writing it (Hyks 2005: 7). However, she admits that it is possible that in the future translators and audio describers will collaborate (Hyks 2005: 7). Unfortunately, the opinion on the strategy of script translation as a time-consuming process – although it is undoubtedly interesting, as an opinion of a practitioner – is not supported by any research results and it is only a personal judgment of the above quoted publication's author.

The topic of the possibility of translating audio description scripts for films was also investigated by a Spanish scientist Juan Francisco López Vera from the Autonomous University of Barcelona (UAB). In an article titled *Translation of audio description scripts – the way forward? Tentative first stage project results* (2006) he presents the outcomes of the first part of the experiment he carried out. This article is a summary of a conference speech which was given during the MUTRA – Audiovisual Translation Scenarios conference. López Vera assumed that the process of creating audio description is too time and money consuming and that the translation of scripts could be an alternative solution that would help to overcome these obstacles and ensure greater access to audio described films (2006: 3–5). To prove his hypothesis, he decided to conduct an experiment comprising of five parts, although he managed to carry out only one of them: a comparison of the time of writing an audio description script and the time of translating the script from English to Spanish. The research was not continued and its author, who probably was a PhD candidate at UAB, apparently quit a scientific career, since none of his further publications were found. López Vera invited four audiovisual translators to participate in his experiment (two dubbing dialogists and two subtitlers) as well as an academic teacher, a dubbing director and an editor[32]. The participants of the research had to:

- **Task one:** watch two films in original version, prepare audio description scripts for the first ten minutes of the original version of each film, adjust the audio description scripts to the Spanish dubbing versions of both films, and read both scripts aloud in order to check if the prepared texts fit in the space between the dialogues;
- **Task two:** watch two films in original version, translate the audio description scripts for the first ten minutes of both films from English to Spanish; adjust the translated scripts to the Spanish dubbing version of both films and read the scripts aloud in order to check if the prepared texts fit in the spaces between the dialogues (López Vera 2006: 7).

The experiment's participants were asked to measure and write down the time they needed to fulfill the tasks. The results helped López Vera to count that in order to prepare audio description for a 90 minutes film one needs circa ten hours and a half, whereas the translation of audio description takes a little more than ten hours. On the one hand, the researcher proved that translation of audio description scripts is a little less time-consuming process, but on the other hand,

---

32 A person who edits the already translated text of an audiovisual translation.

one has to admit that this difference is so small that it seems to be insignificant. What is more, it is hard to consider these results as measurable, because, as the author underlines himself, these are only initial results, which were prepared on the basis of the first of the five parts of the research, and the work of first ten minutes of the film. Another aspect that raises doubts is the way of selecting the participants – it is not clear whether they were trained in audio description, what was the quality of the description they wrote, and whether only translators were asked to translate the scripts, or did people who were not necessarily competent in this matter, like a director or an editor, also participate. In the latter case, the people's involvement it the translating process could make it last longer and influence the quality of the final product. Unfortunately, as it was mentioned before, the research was not finished and Juan Francisco López Vera did not publish more texts on this or any other topic.

The question of translating scripts was also investigated by Dolores Herrador Molina in her MA dissertation *Translation audio description scripts from English to Spanish*[33] (2006), which she wrote at the University of Granada. The author's aim was to determine whether a translation of scripts form English to Spanish is feasible, what problems can a potential translator encounter and whether the possible audience of audio description, accustomed to the Spanish style of this kind of texts, will accept the proposed solution. In order to check the feasibility of translating scripts from English to Spanish, the author translated three fragments of the English audio description for a film called *The Hours* (dir. Stephen Daldry, 2002).

After conducting an analysis of the translating process, the author of the thesis came to a conclusion that translating scripts from English to Spanish can cause problems, since the English script has a tendency to include: redundant information, accumulation of adjectives and subordinate clauses, frequent use of possessive pronouns and idiomatic expressions (Herrador Molina 2006: 35–41). However, it turned out that these problems can be solved by the use of following strategies: omission of redundancy, elimination of some adjectives, avoidance of excessive use of possessive pronouns or using them instead of articles, change of subordinate clauses into coordinate clauses, use of Spanish equivalents of the idiomatic expressions or applying the explicitation strategy (Herrador Molina 2006: 42–44). In conclusion, one can say that the problems a translator can encounter when translating an audio description script from English to Spanish,

---

33 *La traducción de guiones de audiodescripción del inglés al español.*

as well as from every other foreign language, are not different than the ones encountered by literary or specialized texts translator.

Then, the fragments translated by the author were recorded. During the recording process the author discovered that after reading, the translation does not fit in the free spaces between the dialogues, although the amount of words is smaller than in the original English audio description. However, it turned out that it is not necessary to shorten the script, because when the voice talent started to read faster and with a more vivid intonation than the one typical for reading audio description in Spain, the script fit in the available space (Herrador Molina 2006: 44–46).

In the second part of the research the author presented three fragments of films with the translated audio description she prepared, and the equivalent fragments with Spanish audio description prepared by the National Organization of Spanish Blind People (ONCE) to a focus group of four blind people from 22 to 43 years old (Herrador Molina 2006: 59). Although the rather small size of the focus groups can raise doubts, one should take into consideration that gathering a bigger group is often a problem for researchers collaborating with visually impaired people. After watching six fragments of the film, the participants of the research were asked to answer the previously prepared questionnaire. The results of the research proved that the Spanish audience positively received both the translated and the written version. It dispelled doubts that audio description scripts translated from English could be rejected due to a stylistic dissimilarity of the English and the Spanish scripts, which – as the author informs – are shorter and less flowery (2006: 71–72).

Almost simultaneously two researchers from the same university, Julian Bourne and Catalina Jimenez Hurtado, began a research whose results were published in the article *From the visual to the verbal in two languages: a contrastive analysis of the audio description of "The Hours" in English and Spanish* (2007). Their research, which was based on a comparative analysis of language structures (including the use of verbs, adjectives, adverbs and syntax) led them to believe that translation of scripts is worth further investigation. However, they pointed out that a potential translation would demand adjusting the text not only to the stylistic conventions of a given language, but also to the rules of audio description characteristic for a given country. This concern, however, seems to be unjustified in the context of Herrador Molina's research and the lack of empirically supported information on the target audience's attitude to the existing standards of audio description, emphasized by Bourne and Jimenez Hurtado themselves.

Very interesting information is provided by Yota Georgakopoulou (2009), who discusses the first attempts of introducing audio description in Greece in her article entitled *Developing Audio Description in Greece*. The author of the text, who was then the chief executive of the European Captioning Institute[34] and today is a manager of research and development of the Deluxe Digital Studios[35], e.g. discusses the question of the new audio describers' training. She claims that the first step of the trainees was getting to know the English audio description scripts – its purpose was to learn the applied strategies of content selection (Georgakopoulou 2009: 40). The next step was writing audio description for a couple of episodes of Greek programs by the specialists from the British and American branch of the European Captioning Institute (Georgakopoulou 2009: 40). Professional audio describers were given audiovisual materials in Greek, together with notes that described the plot in English (Georgakopoulou 2009: 40). On this basis they wrote scripts in English, which then were sent to Greece, where they were translated into Greek and adjusted to the Greek standards by professional audio describers (Georgakopoulou 2009: 40). This way several programs were prepared and then the Greek audio describers were assumed to gain enough experience to write scripts on their own (Georgakopoulou 2009: 40). During the training, several interesting observations were made:

- Since English is more precise and has broader lexis than Greek, and due to some time and space restrictions of audio description, the loss of some content is inevitable.
- Greek language convention demands that a text to be read aloud should have short sentences. Therefore a single sentence of Greek audio description includes less information than the syntactically complex English sentences.
- When translating script from one language to another, one should also take into consideration the non-linguistic questions, like the knowledge and expectations of the audience, which are different, especially when it comes to culture-related elements (Georgakopoulou 2009: 40).

The possibility of translating audio description scripts was also taken into consideration at the margin of the European research project *Digital Television for All* conducted between 2009 and 2010 by Pilar Orero from the Autonomous University in Barcelona. Unfortunately, the final results of the research are not connected with the issue of scripts translation (Mazur & Chmiel 2012).

---

34 Audiovisual translation provider.
35 Audiovisual translation provider.

The option of translating audio description scripts from English to Polish for dubbed films was also considered by Agata Psiuk in her MA dissertation written at the UNESCO Chair for Translation Studies and Intercultural Communication of the Jagiellonian University. This thesis was a part of the project, which resulted in a PhD dissertation on which this book is based. For the purpose of her analysis the author wrote an audio description script and translated audio description scripts for the film *Harry Potter and the Prisoner of Azkaban* (dir. Alfonso Cuarón Orozco, 2004) from English to Polish. The author wrote the audio description script in 70 hours and 45 minutes, including time synchronisation, while translation of the script together with an adaptation and time synchronisation took 22 hours (Psiuk 2010: 71). When quoting these calculations one should remember that it was the first script written by this author. Therefore, on one hand it can be assumed that in the course of time the script writing process can be reduced. On the other hand, one should also emphasise that the script written by Psiuk was consulted neither with a blind person nor a more experienced audio describer – which is necessary when it comes to beginners. Therefore, one should add the time which is necessary for a consultation to the time calculated by the author. When it comes to translating audio description such a consultation may be useful as well. Since it is a translation of a script prepared by a professional author, one can assume that the number of corrections would be smaller and it would only refer to the language level, and not the choice of the content the scripts convey.

A comparative analysis of both written and translated types of scripts proved that the scripts are different when it comes to the language form, but they do not differ significantly when it comes to the content they convey. The author states that:

> If the translation of the script constructs the same content which is present in the audio description written by an audio describer, using only a different set of tools, it means that the overarching objective of audio description can be achieved. A blind person who watches a film with audio description will reconstruct the mental images encoded in the linguistic picture of the world in both cases: when the audio description for the film will be prepared on the basis of writing a script, and when the script will be a translation of a script, which was written previously in a different language (Psiuk 2010: 97–98).

Apparently, the topic of audio description scripts translation generates interest of both professionals and students, although at this moment this interest is not great. The available publications are certainly a step towards investigating the possibility of translating audio description scripts. Unfortunately, due to a limited research scope they were based on, they do not provide enough arguments allowing the evaluation of how realistic it is to apply such a solution.

4. Research methodology

In order to answer the question of the feasibility of creating audio description through translation strategy, a research comprising of three research experiments was conducted. The purpose of the first experiment – a **time-consumption analysis** – was to determine the average time needed for writing an audio description script and translating it from English to Polish. The participants of this experiment were the students of the UNESCO Chair for Translation Studies and Intercultural Communication at JU, the students of the Institute of English Studies at JU and the graduates of master studies of audiovisual translation at the Imperial College in London.

It was written in the introduction that science should not serve for science exclusively, but also for the good of a man. According to the rules of *action research* it is assumed that the members of a given community have the greatest competence to define what is most important for them. Therefore, their participation in a search for the practical solution of a given problem is inevitable, and this search should acquire a form of evaluation and reflection, which help to introduce practical changes and improvements. (Surdyk 2006: 913). Taking this into consideration, the **second experiment** was conducted. **A pilot study**, whose purpose was to investigate the reaction of the visually impaired viewers to the proposed solution, namely the strategy of translating audio description scripts. For this purpose a pilot study was conducted with both an experimental group of visually impaired youth, and a control group of young people without vision impairment. Both groups were presented with fragments of audio description scripts fragments of an audio description created as a result of applying the strategy of translating and writing, and then they were asked to choose the description they preferred. The purpose of the **third experiment**, which was **the cognitive comparative analysis** of three audio description scripts written by novice audio describers and three scripts translated by novice translators, was to check how those texts differ and find the reasons of possible causes for the preferences of the experimental and the control group.

5. Time-consumption analysis

The former research results (López Vera 2006 and Psiuk 2010) did not give an unambiguous answer to the question of the lesser time consumption of the translation strategy. In López Vera's research, the difference between the process of writing audio description and translating it was only half an hour.

However, it seems that this research is not credible due to a couple of factors. Firstly, the conclusions were based on the analysis of data compiled on the basis of the results referring to the first ten minutes of a film only. Secondly, as it was already mentioned before, the subsequent stages of the writing and translating process investigation raise many doubts. Thirdly and most importantly, the research has never been completed. It is true that Psiuk proved the significant difference between the time needed to write an audio description script and to translate it from English to Polish, but the limited scope of this research does not allow to draw binding conclusions and state that translating audio description scripts from English to Polish is less time-consuming than writing them originally in Polish. Other research on translating audio description scripts did not take the time factor into consideration. Therefore, it seemed necessary to conduct a detailed measure of time needed to write a script and the time needed to translate it.

5.1 Research procedure

In order to collect the data needed for analysis a research experiment was conducted. The participants of the experiment were: the students of MA studies at the UNESCO Chair for Translation Studies and Intercultural Communication, the students of the Institute of English Studies at the JU and four Spanish-speaking graduates of the post-graduate studies of audiovisual translation at the Imperial College London.

The purpose of this book is to test the possibility of applying the strategy of translating audio description scripts from English to Polish for the foreign-language films dubbed in Polish. However, the number of students interested in this research was so big that it enabled to collect the data which exceeded the research goals. Therefore, as a part of the experiment, the time-consumption of translating the scripts from English to Spanish to the English-language films with Spanish dubbing was also tested, as well as the time-consumption of the process of writing and translating audio description for English spoken films with a Polish voice-over. At the beginning of the research, the use of audio description in the films with voice-over was considered as impossible. However, since the research in this matter was planned and the students showed great enthusiasm, it seemed proper to also verify the possibility of applying the strategy of writing and translating scripts from English to Polish to English spoken films with a Polish voice-over. Today, when the usefulness of creating audio description for foreign-language films with a voice-over has already been confirmed (Wilgucka 2012, Szarkowska & Jankowska 2012 and

Jankowska, Szarkowska & Wilgucka 2014), the decision of making effort to verify the time-consumption of the strategies of translating audio description for those films seems to be completely justified. At this point it is important to notice that some of English audio description scripts which were translated in order to add them to foreign language films with a Polish voice-over, came not only from the UK, but also from the USA. Those scripts, excluding the time-consumption analysis, will not be analyzed for the purpose of this book due to its scope, which is limited to the films dubbed into Polish. However, it is planned to use them in the next research project, whose aim is to verify the possibility of implementing the strategy of translating English audio description scripts in order to use them in foreign language films with a Polish voice-over and audio subtitles. Such a research seems to be really necessary, since the major AVT modes used in Poland are voice-over and subtitles. As a consequence, the increase of the audiovisual culture availability for visually impaired people means also the access to foreign language films without dubbing. One can assume that the results of a research conducted in that matter could be relevant for other subtitling countries who struggle with financing audio description similarly to Poland. As a part of the research it is also planned to conduct a comparative analysis of English scripts from the United Kingdom and the USA, since those two countries seem to differ significantly when it comes to the concept of audio description creation criteria. It is also planned to investigate the target audience's reception, which can possibly answer the question of visually impaired persons' attitude to, e.g. the issue of different styles of audio description.

Before starting the work, the students of the UNESCO Chair were acquainted with basic rules of creating audio description during the class on audiovisual translation taught by the author of this book at the Chair during the summer terms of years 2008/2009 and 2009/2010. The students of the JU Institute of English Studies participated in a special audio description workshop which took place in spring 2010. The graduates of the Imperial College London expressed their interest in taking part in the experiment after they heard a conference presentation on this research and the possibility of translating audio description scripts given by the author of this book in November 2010 during the Languages & The Media conference in Berlin. During their audiovisual translation studies they participated in an audio description class. One can conclude that both novice audio describers and novice translators participated in the experiment.

The students participating in the research were divided into six groups. The **first group** was asked to translate the scripts from English to Polish for following foreign English spoken films dubbed in Polish: *Harry Potter and the Sorcerer's Stone* (dir. Chris Columbus, 2001), *Harry Potter and the Prisoner of Azkaban* (dir. Alfonso Cuarón Orozco, 2004) and *Ice Age 2: The Meltdown* (dir. Carlos Saldanha, 2006). There were two reasons for choosing those particular films. Firstly, Krzysztof Szubzda gave us an access to audio description scripts he created for the films *Harry Potter and the Sorcerer's Stone* and *Ice Age 2: The Meltdown*, which he created at the beginning of his audio describing career, after an initial training in London. Therefore they could have been treated as scripts written by a novice audio describer and used in the cognitive comparative analysis of scripts written by novice audio describers and translated by novice translators, whose results will be described in a later part of this book. To complete the data needed for the analysis, three translated scripts and one written script were needed. Additional argument in favor of selecting those three films was that we acquired original English versions of the audio description scripts. The **first group** comprised of three female students. They translated audio description scripts for each of the films. The task of the **second group** was to create audio description scripts for the above mentioned films. Since Krzysztof Szubzda offered two scripts – for the films *Harry Potter and the Sorcerer's Stone* and *Ice Age 2: The Meltdown* – it was only necessary to write a script for the film *Harry Potter and the Prisoner of Azkaban*. There were three female students participating in the second group and the effects of their work were two audio description scripts for the film *Ice Age 2: The Meltdown* and one script for the film *Harry Potter and the Prisoner of Azkaban*. The scripts for *Ice Age* written by the students were included only in the time-consumption analysis. In the comparative analysis, the script written by Krzysztof Szubzda will be used. The scripts to *Ice Age* will be used in another research project, whose aim is to analyze the mistakes made by novice audio describers. **The third and fourth group**, including ten people altogether, were assigned the work on foreign language films with Polish voice-over. They worked on following films: *Iris* (dir. Richard Eyre, 2001), *Big Fish* (dir. Tim Burton, 2003), *The Birdcage* (dir. Mike Nichols, 1996), *Mission Impossible* (dir. Brian de Palma, 1996) and *The Shawshank Redemption* (dir. Frank Darabont, 1994). The films selection criterion was to represent a possibly broad spectrum of film genres. The fact that the author of this book managed to acquire original English audio description scripts for all the films mentioned above, written by audio describers from the UK and the USA, was also significant. **The third group** was asked to translate audio description scripts from English to Polish, and **the fourth group** was asked to write scripts to above mentioned films. These two groups managed to produce ten scripts: five translated and five written ones. The members of the **fifth and sixth** group were four Spanish speaking female students, who translated and created audio description scripts from English to Spanish for the film *Harry Potter and the Sorcerer's Stone* in its Spanish dubbing version. They created two Spanish scripts and two translations of English audio description scripts to Spanish.

The task of all students participating in the experiment was to deliver a final audio description script which was ready for reading. It means that the scripts had to be not only written or translated, but also adjusted to the space available between the dialogues, including time codes. All participants of the research were asked to measure and note the time needed to fulfill the tasks they were given. Concerning the scope and goal of the tasks, it was not necessary to measure the time of fulfilling particular stages of the process of creating an audio description script. Therefore the participants were asked to provide the whole data concerning the time needed to write or translate a script, from the moment of beginning the work on creating a script to the moment of its finishing, excluding the time necessary for getting to know the film.

The scripts created by the Polish students as a result of applying the strategy of writing and translating were proofread. The scripts created in Spanish were not proofread, due to the impossibility of finding a proofreader who meets both language and substantive requirements, which are not met by the author of this book, although she graduated from Spanish studies. Therefore the results pertaining to the time-consumption of the strategy of writing and translating in Spanish were not presented in the final data summary.

5.2 Results of the time-consumption analysis

Below you can find the data collected during the research experiment aimed at measuring the time-consumption of two strategies of creating audio description, namely writing and translating.

Let us start with creation of audio description scripts for foreign productions dubbed in Polish. One can conclude that the average time of writing an audio description script is 45 hours, and the time of translating a script from English to Polish – approximately 15 hours. Interestingly, there is no visible relation between the duration of the film and the time needed to write and translate audio description. However, it is not surprising – experienced audio describers, when asked how much time they need to create audio description to a 100 min. film, emphasized that the time needed to create a script depends on the characteristics of a particular film being worked on, and not on its duration only.

Table 1: Time of writing and translating AD scripts for films dubbed in Polish

Title of a film	Film's duration	Writing	Translating
Ice Age 2: The Meltdown	86 min.	35 h	12 h
Harry Potter and the Sorcerer's Stone	147 min.	30 h	14 h
Harry Potter and the Prisoner of Azkaban	147 min.	71 h	20 h
Average time	126 min.	45 h	15 h

Table 1 presents an interesting phenomenon, which is the big time divergence in case of the written audio description – one of the authors spent twice as much time on creating the script than the remaining two. However, at this point it is necessary to explain some personal matters, since the factor determining this difference also belongs to them. When working with students, the author of this book could get to know their method and quality of work quite well. The student who prepared an audio description script to the film *Harry Potter and the Prisoner of Azkaban* is a very scrupulous and diligent person. It is true that she needed almost twice as much time than the others to create the script, however, maybe this is why her script – unlike the remaining two – did not require proofreading, which is described below in greater detail. It seems that such a high quality of a script could have been influenced by the fact that its author translated the script from English to Polish several months before – as a part of preparation for writing her MA dissertation. It can mean that the strategy of script translation turned out to be a useful tool when it comes to teaching audio description.

When it comes to the results of the experiment whose participants were the Spanish speaking graduates of the post-graduate studies of AVT at Imperial College in London, one can classify them as the results of the control group. The results presented below proved that an average time of translating an audio description script from English to Spanish is 17.5 hours and writing the script in Spanish took circa 32.5 hours. All the authors and translators from Spain worked on the scripts for the film *Harry Potter and the Sorcerer's Stone*. The fact that the participants of the research, who worked on the same film, needed different amount of time to write or translate a script, allows to conclude that the question of time-consumption of the process of creating a script also certainly depends on individual skills or talents of the audio describers and translators.

Table 2: *The time of writing and translating AD scripts for films dubbed in Spanish*

Title of a film	Film's duration	Writing	Translating
Harry Potter and the Sorcerer's Stone	147 min.	30 h	20 h
Harry Potter and the Sorcerer's Stone	147 min.	35 h	15 h
Average time	**147 min.**	**32,5 h**	**17,5 h**

The results presented in Table 3 prove that when it comes to the work on audio description scripts for English language films with a Polish voice-over, the average time needed to write a script is 27 hours, and to translate it – circa 14 hours. Similarly to other examples it does not seem that the duration of a film influences the amount of time needed to prepare a script, either by writing it or translating it from English to Polish.

Table 3: *The time of writing and translating AD scripts for films with Polish voice-over*

Title of a film	Film's duration	Writing	Translation
Big Fish	120 min.	24 h	12 h
Iris	87 min.	24 h	24 h
The Birdcage	119 min.	35 h	10 h
The Shawshank Redemption	142 min.	22 h	13 h
Mission Impossible	105 min.	30 h	12 h
Average time	**115 min.**	**27 h**	**14 h**

The scripts prepared by the students – both the translated and written ones – required proofreading. When it comes to the translated texts, quantity corrections were needed –mostly related to the necessity of adjusting the length of the text to the space available between dialogues and it was limited to an insignificant shortening of the text, e.g. by founding a shorter synonym. It was not necessary to remove entire fragments of the text. Unfortunately, when it comes to the written texts, it was necessary to revise them thoroughly, both regarding the quantity and quality. The average time needed for the correction of the translated scripts was circa 8 hours, while 20 hours when proofreading the written texts.

In Table 4 provided below you can find average times of writing and translating audio description scripts for dubbed films and films with voice-over, including the time spent on proofreading. They prove that the average time of creating a script when applying the strategy of writing is circa 56 hours, and when applying the translation strategy – circa 22.5 hours. It is worth noticing that the time of writing an audio description script does not take into consideration the time needed for consultations with a blind person, and the participants of the research were not asked to hold them. On the basis of the data provided by experienced audio describers (see 5.3) this time should be estimated as 6 additional hours. In case of applying the strategy of translating a script from English, such consultations are not necessary, since they were already conducted at the stage of the text preparation in the source language – therefore it is assumed that the script is ready after a translation, which also includes a time synchronization.

Table 4: The time of writing and translating AD scripts to a Polish-language production, including proofreading

Writing		Translation	
Writing	Proofreading	Translation	Proofreading
36 h	20 h	14.5 h	8 h
56 h		22.5 h	

5.3 Time-consumption analysis: summary and conclusions

The time-consumption analysis proved that in case of the films dubbed into Polish, the translation of the scripts from English to Polish took three times less than writing the script in one's mother language. The results of the control group comprised of the native speakers of Spanish translating audio description scripts from English to Spanish and writing a script in this language are slightly different. In this case, the translation of the text required twice less time. This can be attributed to a number of reasons – from the differences between languages to the experience and skills of both those translating and those writing the scripts.

The analysis showed that also in case of foreign language films with a Polish voice-over the translation of scripts from English to Polish requires twice less time than writing them.

After adding to the above mentioned times the time needed for proofreading, and – in case of the written scripts – time for consultation with a blind person and the second audio describer, it turns out that preparing the final version of the written script takes almost three times longer than preparing the final version of a translated text. It is worth comparing the data obtained in the research experiment with those delivered by experienced audio describers who were asked about the average time they needed to prepare audio description for a 100 minutes long film. The results, collected via e-mail or during conversations, are presented in Table 5.

Table 5: Complete time of preparing an audio description script by experienced audio describers

Audio describer	Writing	Self-correction	Consultation with another audio describer	Consultation with a blind person
1	30	4	6	5
2	40	8	0	7
3	35	8	0	5
4	12	3	0	5
5	40	8	0	7
6	30	5	0	6
7	40	8	8	8
8	20	5	0	8
Average	31	6	2	6
Total average	**45 h**			

The above data show that experienced audio describers need on average 45 hours to prepare a final version of an audio description script for a 100 minutes long film.

The conducted research experiment proved that the strategy of creating an audio description script through translation is three times less time-consuming than the strategy of writing, both when it is applied by novice translators and novice audio describers. What is more, the strategy of creating an audio description script through translation from English to Polish is twice less time-consuming than the strategy of writing, even in case when the translation is done by a novice translator and the author of the audio description script is an experienced audio describer. Taking into consideration these results, it is worth considering conducting a research whose aim would be to check the time consumption of the strategy of creating the scripts through translation done by professional translators. One can assume that the difference between the time consumption of the translating strategy and the writing strategy would be even bigger. Therefore it can be estimated that if blind people accept scripts created as a result of the translation strategy, it would turn out that this strategy is not only less time-consuming, and therefore more economical, but also appropriate when it comes to quality than would enable to **put it into practice.**

6. Pilot study

As it was described in a greater detail in chapter 3, until today the question of the possibility of creating audio description scripts through an application of the strategy of translating them to one's mother language was not a subject of many research studies. The conducted studies were mostly focused on the aspect of this process' time-consumption and to some extent on the question of possible translation difficulties, and their results gave only partial answers to the stated questions.

Only one of the already conducted research studies on the possibility of translating audio description scripts included the investigation of the reception of such a solution by the target audience. Unfortunately, there were only four participants of this study, which is not a sufficient research sample. The aim of this book is to work out a practical solution, and who, if not the members of a given community, are the most competent ones to decide whether it is acceptable?

The detailed results of the research are presented below. In the first part the data collected during the experimental group study are discussed and afterward the control group's results. In the last part of this chapter, the results of both groups' results are confronted and then summarized.

6.1 Research procedure

The pilot study had two phases. The first was the **experimental group study,** which was conducted on the 14th of January 2010 in the School and Education Centre for Blind and Visually Impaired Children in Kraków. The data for the **control group** were collected in two ways: through an Internet questionnaire and during a survey conducted in Henryk Sławik Junior High School no. 19 in Katowice.

6.1.1 Research procedure in the experimental group

The participants of the research were asked to watch the film *Harry Potter and the Sorcerer's Stone* with audio description and answer the questionnaire. The film presented to the audience was prepared in a special way. Translated audio description was added to the first part of the film. It was created by Diane Langfold and translated to Polish by the then student of the UNESCO Chair, Justyna Drożdż-Kubik. In the second part of the film, a different audio description in Polish was used – the one created by Krzysztof Szubzda, the first Polish audio describer. Both audio descriptions were read by the same voice of a speech synthesizer. The participants of the research were not informed that audio descriptions of two different authors were used, so that their reception of the film was not influenced and the answers were not suggested in any way. Such a preparation of the audiovisual material was aimed at investigating the target audience's reaction to a script translated from English and a script that was written in Polish. This reaction is of crucial significance for the results of the study presented in this book.

6.1.1.1 Preparation of the film

The audio description was provided with sound through a text-to-speech software (TTS), following the method proposed by Szarkowska (2011: 144). We have used a TTS Ivona, which was lent for this purpose by its producer – IVO Software company. During the process of preparing the audio description script to be read by the synthesizer it turned out that Szarkowska's proposal demands some modifications. In the above mentioned method the speed and volume of reading are set in the speech synthesis program. However, such a setting has a disadvantage: both the volume and the speed levels are the same through the entire film. In some cases it meant that during loud scenes the audio description was not audible, and during silent scenes the sound of audio description was certainly too loud. A similar problem occurred when it came to the speed of reading. Setting the speed to medium eliminated the possibility of, e.g. faster reading of the text during the short brakes between dialogues. Therefore – after consulting the speech synthesizer's producer – HTML codes (e.g. <prosody volume="x-loud">) were added to the script, which allowed increasing or decreasing of the speed and volume of some utterances. Additionally, taking into consideration the speech synthesizer's specific way of pronunciation, the foreign word's notation was changed to a phonetic one (e.g. Hary instead of Harry or Lokhart instead of Lockhart, the punctuation was modified (e.g. commas were removed in order to eliminate a long pause in reading) and the notation of some words was adjusted, so that the reader could pronounce them properly. All the modifications mentioned above are visible in Example 3.

Example 3: Audio description script to be read by a speech synthesizer (author: Justyna Drożdż-Kubik).

01:54:38 Hary odpycha czarodzieja i wymierza w jego kierunku różdżkę.

01:54:55 <prosody volume="x-loud">Ron strąca Lokharta do dziury.<prosody>

01:55:23 <prosody volume="x-loud">Hary wskakuje do otworu, za nim Ron.<prosody>

01:55:37 <prosody volume="x-loud">Hary i Ron lądują na kościach. Tuż obok Lokharta.<prosody>

01:55:55 Hary wbiega do jednego z kanałów, za nim Lokhart i Ron.

01:56:02 Przechodzą do niskiego pomieszczenia. Wszędzie leżą zeschnięte skorupy.

01:56:21 Lokhart mdleje

In the method proposed by Szarkowska a specially prepared audio description script is read by the TTS simultaneously with the film. In order to prepare the materials to the pilot study a different technique was used. To avoid possible technical errors (e.g. the speech synthesis software freezing), the synthesizer's reading was recorded as a WAV sound file and then permanently connected to the film by Sony Vegas Pro – software used for editing video files. This software enabled also precise adjusting of the audio description's track's volume to the film's sound. The final effect was a film with an additional audio description track audible in the generally accessible sound channel.

6.1.1.2 The questionnaire

The questionnaire filled in by the participants of the survey was prepared on the basis of exemplary questionnaires available at RNIB and provided by researchers who investigate the topic of audio description: Agnieszka Szarkowska from the Institute of Applied Linguistics at the University of Warsaw, as well as Agnieszka Chmiel and Iwona Mazur at the Adam Mickiewicz University in Poznań.

The questionnaire was comprised of eleven general and eight detailed questions (see Appendix 1). The general part included only multiple choice questions, while the detailed part included two open questions. The ratio of multiple choice questions to the open questions was the result of the conditions in which the survey was conducted. The participants of the survey, who were visually impaired, needed help of the researchers who conducted the survey and the workers of the centre, who filled the questionnaire individually with every participant. The prevalence of multiple-choice questions helped to facilitate this process. The participants answered the general questions before the film's screening, while the detailed questions were asked after the end of the film.

The aim of the eleven general questions (see Appendix 1, questions 1–11) was to identify the age of the respondents, their gender, education, the moment of becoming blind and the degree of vision impairment, their customs concerning watching audiovisual productions and the experience in watching productions with audio description. When it comes to determining the respondents' age, the division into groups proposed in the questionnaire was based on Lipińska's proposition (Lipińska 2006: 57–67). Lipińska's method – in accordance with the stages of human development presented by Erickson[36] – takes into consideration such parameters as an ability to analyse and connect facts as well as abstract thinking, which seems to be particularly important for the reception of a film with audio description.

Next two questions (see Appendix 1, questions 12–13) referred to the difference between two halves of the film. The participants were asked if and what kind of difference they noticed between the two halves of the film. As it was already mentioned, the first half of the film *Harry Potter and the Sorcerer's Stone*, which was displayed during the study, included the translated audio description, whereas the second half included the written audio description. In order not to influence the opinion of the respondents, they were not informed about this difference and both audio description scripts were read by the same synthetic voice.

In the subsequent detailed questions (see Appendix 1, questions 13–19) the participants of the study were asked to point out the preferred fragments of the description. To investigate if the participants of the research preferred the written or translated audio description, they were presented two fragments of audio description for the same scene in the film. One of the fragments was the audio description created through the strategy of writing and the second one – through the strategy of translating. There were 6 scenes selected altogether. The sequence

---

36 Children: infant stage (up to 8 months of life), young childhood stage (up to 3 years old) and the age of play (3–6 years old). Youth: school age (6–12 years old) and adolescence (12–18 years old). Adults: early adulthood (18–35 years old), middle adulthood (35 years old to retirement) and the retirement period.

of presenting the descriptions was random. While in one question the translated audio description was at the first place, in the next question it was situated at the second or the first place. This procedure was applied in order to avoid creation of a scheme that would suggest the respondents which of the descriptions was written and which one was translated.

The answers to all the questions discussed above will be presented in the further part of this book (see 6.2). At this point it is also worth noting that the questionnaire included questions about the customs and needs concerning foreign language audiovisual productions with a voice-over. The answers to these questions will be omitted in this book, since its scope is limited to dubbed productions. However, the enthusiastic answers of the respondents became an inspiration to start a new research project, whose aim is to check the possibility of using audio description in voiced-over films. The answers also confirmed that it makes sense to check the possibility of transiting audio description scripts for foreign films with Polish voice-over.

6.1.1.3 Selection and collection of data

As it was already mentioned, the data were collected during a special screening of the *Harry Potter and the Sorcerer's Stone*, which was organized thanks to the agreement and help of the teachers and carers from the School and Education Centre for Blind and Visually Impaired Children. The screening took place at the time of regular lessons and the participants were twenty two children from 6 to 18 years old. The age of the participants, who belonged to the group of school age or adolescence, is justified by the specificity of the film chosen for the pilot study, as well as of all the films that were analyzed for the purpose of this research. It was also expected that the opinions of the survey's participants would be influenced by their ability to analyze and connect facts as well as abstract thinking, which is different for people of ages 6–12 and 12–18.

Before the film was displayed, the participants of the research answered the general part of the questionnaire. After the projection they answered the detailed questions. The participants of the research were assisted by the teachers and carers of the centre, as well as by UNESCO Chair students: Agata Psiuk and Justyna Drożdż-Kubik and the author of this book, who helped them to write down the answers.

Unfortunately, after an initial analysis of the collected research material, five questionnaires had to be rejected, since they were incomplete – the respondents did not answer the questions crucial for the purpose of the research, pertaining to their preferences concerning audio description created as a result of applying either the strategy of writing or translating a script.

After the second analysis of the material was conducted, one more question-naire was rejected – the one answered by a respondent from the 6–12 years age group. The reason of excluding the above mentioned person from the re-search sample was the fact that this participant was the only representative of the above mentioned age group, and the provided answers could not be con-sidered as representative. Finally, sixteen questionnaires were counted as the research material. They were filled by sixteen participants, including eleven boys and four girls. However, we hope to repeat the reception study with a more diversified age groups.

6.1.2 Research procedure in the control group

As it was already mentioned in the pilot study description, the research was con-ducted with an experimental group, but also with a control group which was com-prised of people without visual impairment. The data were collected via the Internet and directly from the students of the Henryk Sławik Junior High School no. 19 in Katowice, who filled in a questionnaire as a part of a Polish language lesson.

The questionnaire planned for the control group included eight questions (see Appendix 2). The aim of the first two questions was to determine the age and gender of the respondents. The six subsequent questions concerned the control group preferences regarding either translated or written version of an audio de-scription script. In order to determine it, the respondents were presented the same descriptions of the scenes that were previously presented to the experimen-tal group. The participants of the research were asked to write which description of a given scene they prefer and why. Similarly to the experimental group, the control group was not informed about the difference between the descriptions of the same scene. The descriptions were presented randomly, to avoid suggesting an answer. In case of the control group, the film was not presented deliberately, since the viewers could have evaluated a description of a given scene when it comes to its faithfulness to the picture.

The survey was conducted with a group of sixty people between the ages of 6 and 18. However, due to the survey's requirements not all of the answers could be taken into consideration. The final shape of the control group had to be changed so that the only experimental stimulus that made the group different from the experimental group was the lack of visual impairment. Therefore, the answers suitable for the experimental group were randomly selected from all of the gath-ered answers. The data analysis presented below is based on the data collected from sixteen members between the ages of 12 and 18, including eleven male and four female respondents.

6.2 The results of the experimental group research

Below you can see the results of the experimental group research. Firstly, the data presenting the structure of a given group are presented, i.e. gender, the level of visual impairment and the age when the respondent's vision was impaired, television watching customs, as well as the experience of watching films with audio description. Then the results of the questionnaires are presented. The questionnaire's aim was to test whether the respondents prefer the written or the translated audio description. The results were subject to a cross tabulation regarding gender, the level of visual impairment and the age when the vision was impaired, and the experience in watching audio described films, since all of the above mentioned variables can be significant when it comes to the preference for either the written audio description or the translated one. The last part is devoted to a brief discussion of the results obtained in the experimental group.

It should be noted that not all of the respondents answered all of the questions. The results presented below are based on the number of all of the given answers and not on the number of all the people who participated in the research.

6.2.1 Introduction

As it was already mentioned, all of the sixteen participants of the survey were between the ages of 12 and 18. The structure of the respondents' gender presented in Chart 1 looks as follows: women constitute 25% of the respondents, and men 75%.

Chart 1: Gender structure of the experimental group.

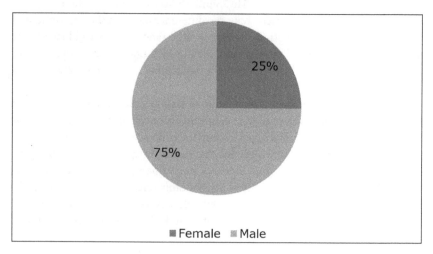

On the basis of the respondents' answers the degree of visual impairment was determined, as well as the moment when a given person lost their vision. Both variables are presented in two subsequent charts. The collected data proved that 12,5% of the interviewees suffered from profound vision loss, 25% – severe one, 37,5% – moderate one and 25% – mild one. According to the World Health Organisation's definition, 37% of the research participants are blind and 63% are partially sighted.

Chart 2: Vision loss degree.

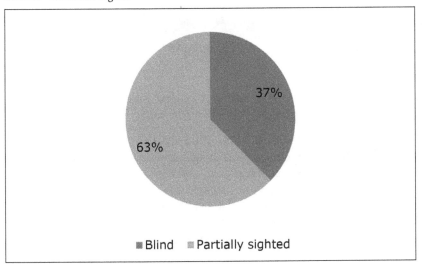

As it is shown in Chart 3 the major part of the respondents, which is 82%, declares that they do not see since they were born, whereas a vision loss after the age of 4 is confirmed by 18% of the respondents. It is, however, worth noticing that in the group of blind participants, 75% of the interviewees does not see since they were born and 25% lost their vision after the age of 4. When it comes to the group of partially sighted participants, the 86% of the respondents declared vision impairment since they were born, whereas 14% answered that their visual impairment happened not earlier than at the age of four.

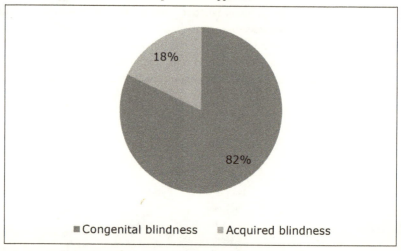

According to Chart 4 the participants of the research differ significantly when it comes to their daily television watching customs. A total of 38% of the respondents belong to two extreme groups – of those who spend the smallest and the biggest amount of time on watching television, 19% per each group. 31% of respondents belong to the groups that daily spend from 1 to 2 and from 2 to 3 hours on watching television. Interestingly, no significant differences were determined between blind and partially sighted people in this matter.

Chart 4: *Audiovisual productions watching customs.*

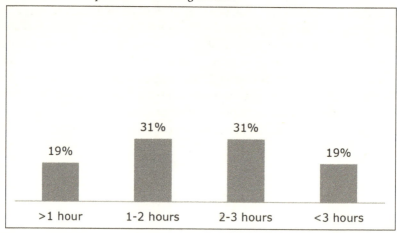

The young people participating in the research were also asked about their experiences in watching films with audio description. As it was expected, only a small group of them had an opportunity to watch a film with audio description earlier. The chart provided below proves that 12.5% of the research participants saw a film with audio description prior to the screening. The majority, 87.5% of the respondents, experienced audio description for the first time during the show organised as a part of the research experiment.

Chart 5: Experience with watching films with audio description.

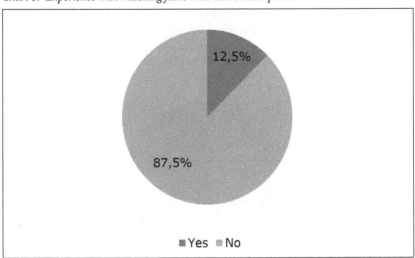

In the next part of this chapter the answers to the questions that allow evaluating the respondents' attitude to written and translated audio description will be presented.

6.2.2 Attitude to written and translated audio description

One of the significant aims of the study was to determine whether the proposed solution, which is the strategy of translating audio description from a foreign language, would be accepted by the target audience of such a product, namely visually impaired people.

In order to answer this question, it seemed necessary to determine whether the visually impaired notice the difference between the written and the translated audio description, what is the nature of the difference and which of the strategies they prefer. Below you will find the respondents' answers to those questions that are related to the issues mentioned above.

6.2.2.1 Seeing the difference – general preferences

After the film's projection the participants were asked whether they saw a difference between the first and the second half of the film. As it was explained before, the interviewees were not informed that one part of the film included translated audio description and the other one – written audio description. Chart 6 proves that the participants' opinions were different. Slight majority of the interviewees declared they did not see any differences between the first and the second half of the film, while 43% said that they saw a difference.

Chart 6: Noticing a difference between two halves of the film.

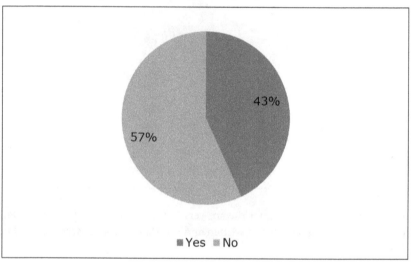

It seems to be interesting that the vast majority (80%) of people who noticed the difference between the two halves of the film are partially sighted. Blind persons constitute only 20% of the group of those who declared that they saw the difference. The fact that among blind persons who answered this question 20% noticed the difference and 80% did not, seems to be even more interesting. Meanwhile in the group of the partially sighted people who answered this question, 56% respondents declared that they noticed the difference and 44% declared that they did not notice it. It is apparent that partially sighted people notice the difference between the scripts more often than blind people.

Chart 7: The vision impairment and noticing the difference between the two halves of the film.

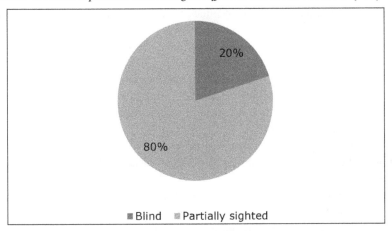

When it comes to the relations between noticing the difference and watching audio-visual productions, it is presented in Chart 8, which proves that 50% of the people who declared that they noticed the difference watch audiovisual productions for 2 to 3 hours a day, 30% – for less than 1 hour and 17% – for 1 to 2 hours. Among the people who did not notice any differences, 25% watch audiovisual productions for 2 to 3 hours a day and also 25% watch them for over 3 hours a day. Therefore, it seems that the amount of time devoted to watching television has no significant influence on noticing the difference between translated and written audio description.

Chart 8: Noticing the difference between the two halves of the film and the customs of watching audiovisual productions.

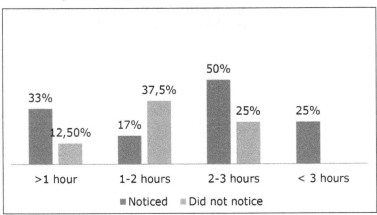

When it comes to the relations between noticing the differences and the moment of losing vision, only people with innate vision impairment declared that they noticed them. Also none of the people who declared their experience in watching audio description did notice the difference between the two halves of the film. There was also no female participant among those who declared that they noticed the difference. However, due to small percentage of people with innate vision impairment, people with experience with watching audio described films and females, those data do not seem to be sufficient to formulate any credible hypothesis that explains them.

The people who declared that they noticed the difference were asked for an additional explanation concerning those differences. Most of the interviewees could not explain what this difference was, but they were convinced that they noticed it. The answers concerned some technical aspects mostly, like the fact that in the first half of the film the volume of audio description was too low in relation to the film, and therefore during some louder scenes the audio description was not audible [the volume was turned up at the participants requests –A.J.], or the fact that in the second half the film stopped a couple of times. Only one of the respondents said that the two halves differ when it comes to the description's details. When asked about explanation, this person said she did not mean that in the first half the description was more detailed, but that "more things were described".

6.2.2.2 Written audio description vs translated audio description – general preferences

In order to get to know the respondents preferences concerning written or translated audio description, they were presented two fragments of audio descriptions prepared for the same scene (see Appendix 1, questions 12–17) and then asked to choose the description they liked more. The respondents' answers are presented in a chart below, which proves that 74% of them prefer translated audio description, whereas 26% prefer the written one.

Chart 9: Experimental group preferences.

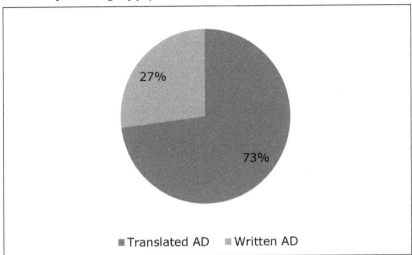

The preferences depending on the gender are presented in Chart 10. Both women and men prefer translated audio description, but some insignificant difference in the preferences is visible – women chose translated audio description more often than men.

Chart 10: Experimental group preferences depending on gender.

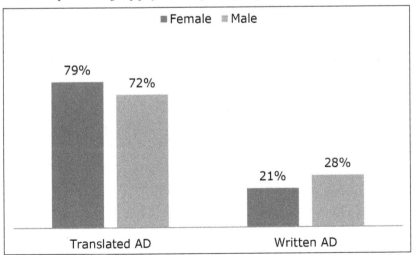

The respondents' answers also prove that there is a difference in the preferences between the blind and the partially sighted, although this difference is slight. As it is presented in Chart 11, blind people preferred the translated audio description more often than partially sighted people.

Chart 11: Preferences of visually impaired respondents, depending on their vision impairment degree.

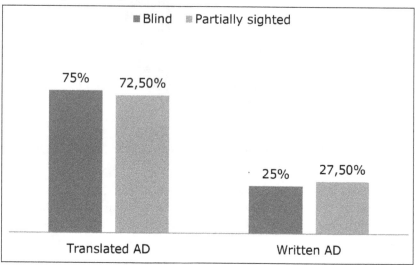

One can also notice that the preference of a given type of audio description is somehow dependent on the moment of vision impairment. According to Chart 12 presented below, the people who were blind from birth preferred the translated audio description more often than those who lost vision between the age of 4 and 9.

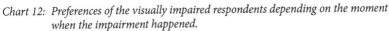

Chart 12: *Preferences of the visually impaired respondents depending on the moment when the impairment happened.*

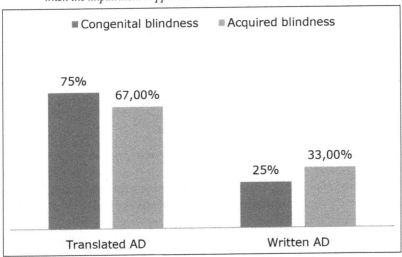

The respondents' preference in relation to particular scenes is worth analysing. Table 6 presented below illustrates the experimental group choices when it comes to preferring the written or the translated type of audio description in particular scenes. The scenes are arranged chronologically: from those in which the written audio description was least popular to those in which it was most often voted as the preferred kind.

Table 6: *Experimental group's preferences in relation to particular scenes*

Scene number	Written audio description	Translated audio description
Scene 5	8%	92%
Scene 6	23%	77%
Scene 1	25%	75%
Scene 3	31%	69%
Scene 4	31%	69%
Scene 2	44%	56%

The data presented in Table 6 suggest that visually impaired persons preferred audio description translated from English in every scene they were presented, however, they did not prefer it with equal enthusiasm in each scene. The translated audio description was preferred by the biggest number of respondents in the scene 5 – in this case the translated description was preferred by 92% of the interviewees. The least popular, almost by half, was the translated audio description of scene 2, which was preferred by 56% of the respondents.

The participants were asked to explain their preferences. Unfortunately, not all of them were able to explain their choices and not all of them commented on each choice. However, those who did, highlighted that the "better" description included more information, was more comprehensible, was not complicated, did not include many details, was more vivid, logical and appealed to imagination better.

6.2.3 Summary of the experimental group results

Summing up the experimental group survey results, one can say that the visually impaired noticed the difference between the two halves of the film to some extent, although most of them could not determine what constituted this difference. They also paid attention to some technical aspects, e. g. the sound volume. One of the participants was an exception and they determined the difference in a surprisingly precise way, noticing that in the first half the audio description included more information, but fewer details.

When analysing the survey's results it is hard to determine clearly whether the fact of noticing the difference could have been influenced by the experience with watching audio described films, gender and the moment of vision loss. The reason for the lack of a clear answer can be the research sample, which was too small, or the fact that the experimental group included too few female participants, those with innate visual impairment and those experienced in watching audio described films. It was not noticed that the amount of time spent on watching audiovisual productions influenced in any way the ability to differentiate the translated and the written audio description. However, it is clear that this difference was noticed more often by partially sighted people than by blind persons.

When it comes to the preferences concerning the written or translated type of audio description, the survey proved that a vast majority of its visually impaired participants prefers translated audio description to the written one, and, what is important, in each of the presented scenes, but with different intensity. Additionally, two minor but real tendencies are visible: it seems that the translated audio description is preferred by females and the blind, whereas written audio description is preferred by males and the partially sighted. This tendency may suggest that the translated and the written scripts differed to the extent which is noticeable depending on gender and the degree of visual impairment.

The respondents were also able to determine the reasons of their preferences – they considered to be better those fragments of the description, which in their opinion included more information, which were more comprehensible, vivid, appealed to imagination and did not include too many details. There were also some arguments emphasising the funny aspect of a given description. The argument was mostly related to the scene 2, which is particularly intriguing, since in this case the written script was preferred by the biggest percent of visually impaired people. The respondents' justifications will be investigated in the part of the book devoted to the comparative analysis of the scripts. They will be quoted in relation to particular scenes.

6.3 Control group survey results

The results of the control group survey are presented below. When it comes to gender and age, its structure is similar to the structure of the experimental group. Control group includes sixteen participants at the age between 12–18, including 25% of female and 75% of male participants. Below you can find the results of the survey whose aim was to check whether the participants prefer the written or the translated audio description. The results were subject to a cross tabulation when it comes to the gender. In the last part the results of the experimental group survey are discussed.

6.3.1 Preferences

As it was mentioned already, the control group was not presented an audio described film. Instead of it the respondents were presented the descriptions of the scenes (see Appendix 2) and asked which of them they preferred and why. The chart presented below illustrates the result of the survey: 56% of the participants without visual impairment preferred the written audio description and 44% – the translated one.

Chart 13: Control group preferences.

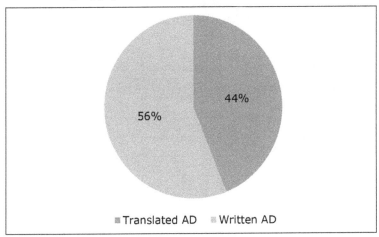

The differences in preferences depending on gender are presented in Chart 14. It is clear that both women and men prefer written audio description. However, a slight difference is visible in this matter. Similarly to the experimental group, women choose translated audio description more often than men.

Chart 14: Control group preferences depending on gender.

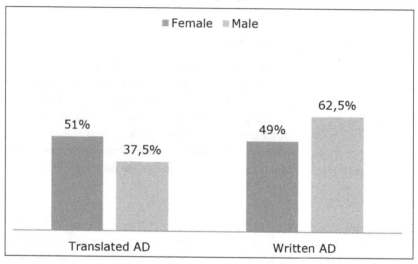

Similarly to the experimental group, it is worth to analyse the preferences of the control group members in relation to particular scenes. Table 7 presented below shows the choices of the control group regarding the written or translated audio description in particular scenes. The scenes are arranged chronologically – from those scenes in which written audio description was least popular to those, in which the biggest percent of respondents preferred this type of audio description.

Table 7: Control group preferences in relation to particular scenes

Scene number	Written audio description	Translated audio description
Scene 3	25%	75%
Scene 1	50%	50%
Scene 5	50%	50%
Scene 6	56%	44%
Scene 4	62,5%	37,5%
Scene 2	81%	19%

According to the Table 7 people without visual impairment preferred written audio description in half of the scenes they were presented, although their enthusiasm was not equal in every case. The biggest number of respondents preferred the written description of scene 2. In two cases the number of voices in favour of both translated and written text was equal and in one case three times more respondents preferred the translated script to the written one.

The participants were asked to justify their preferences. When explaining their choices, they most often referred to the dynamics, emotivity, floridity of the description and its authenticity – it is surprising, but a part of the participants seemed to know Harry Potter's adventures so well that they noticed inaccuracies and false information in comparison with the film (which they did not see during the research) and the book.

6.3.2 Summary of the control group survey's results

To sum up the survey results in the control group one may state that people without visual impairment prefer written audio description. There is also a minor tendency for women to prefer translated audio descriptions. It seems very interesting that people without visual impairment emphasised mostly the importance of the dynamics, details, floridity and emotivity of the description.

6.4 Summary of the results and conclusions

When analysing the compared results of the study presented in Chart 15, one can notice that the visually impaired prefer translated audio description scripts to written ones more often than people without such impairment. However, it turned out that the partially sighted show some tendency to prefer the written scripts.

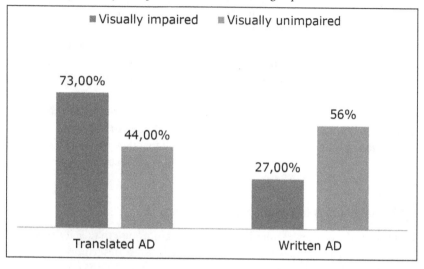

Another interesting aspect is the internal division in the control and experimental groups. As it is shown in Table 8, visually impaired people prefer the translated script in practically every scene they were presented with, although with various levels of enthusiasm, whereas the members of the control group are divided – they generally prefer written audio description, but in some cases more people preferred the translated script or assumed that they are equally good.

Table 8: Comparison of the preferences of the control and experimental group

Experimental group			Control group		
Scene	Written script	Translated script	Scene	Written script	Translated script
5	8%	92%	3	25%	75%
6	23%	77%	1	50%	50%
1	25%	75%	5	50%	50%
4	30%	70%	6	56%	44%
3	32%	68%	4	62,5%	37,5%
2	44%	56%	2	81%	19%

When analysing the above table it is hard to find any relations between the arrangement of the answers of the experimental and the control group. However, it is very intriguing that when it comes to the scene 2, the written script received the highest marks from both of the groups.

The survey proved also that people without visual impairment seem to prefer the descriptions they consider to be more detailed, emotional and colourful. They also paid special attention to the compatibility of a description and a picture, while visually impaired people chose those descriptions that seemed more logical to them, included more information, although not many details, and appealed to their imagination.

One can admit that within the scope designated by this research, the pilot study proved that the target group of audio description would not only be willing to consider translating audio description as a feasible strategy of creating the scripts, but also, in some cases, they would prefer the scripts created in this way.

However, we are aware of the fact that the scope of the study was limited, and the age of the respondents as well as the film's genre can influence their opinions. Therefore, the next step should be a further research on the reception of the films with written and translated audio description, with a more numerous and more diversified experimental group and with different film genres. In connection with the types of audiovisual translation for cinema and television that are adopted in Poland, it seems inevitable to conduct a research on the possibility of translating audio description scripts to foreign films with a Polish voice-over or audio subtitles.

The research confirmed to some extent the truth known in the community of people creating audio description: that blind and visually impaired viewers are not a unified group and that their preferences and expectations concerning audio description differ and can depend on the moment and the degree of their vision loss.

The research we have conducted proved also that gender can be a factor that influences the expectations. Those results can also prove that the process of mental imaging differs depending on the degree of blindness and the moment it occurred. Our research also proved that gender can constitute a diversifying factor. This topic is open for further research of pedagogues working with blind children, psychologists, psycholinguists and neurolinguists.

The explanations concerning the choice of particular description formulated both by the visually impaired and people without vision impairment are also very interesting. When analysing them cross-sectionally it seems that they can be classified into two groups that mirror two functions of texts, namely the poetic and the informative ones. It seems interesting that people without vision impairment paid the most attention to the poetic function mostly, while the visually impaired paid the most attention to the informative function. The discovery that in some cases the poetic function is also important for visually impaired people can also be considered very significant. This thesis is supported by Scene 2: almost half of the visually impaired respondents chose written audio description for this scene, and explained that it was "funny".

It seems particularly interesting from the perspective of this book that both the visually unimpaired and partially sighted people seem to prefer written scripts. This conclusion can suggest that written and translated scripts differ to the extent recognisable by those people and that translated scripts fulfilled the expectations and requirements of blind people in a more satisfactory way than the expectations of partially sighted persons.

To investigate whether the scripts differ and how, and to understand, why visually impaired people unexpectedly considered the poetic function in Scene 2 as important, a cognitive comparative analysis of the scripts was conducted and its results are presented in the next chapter.

7. Comparative analysis of the scripts

The pilot study brought valuable conclusions, but they are also a source of new challenges. They make us consider why the visually impaired preferred the fragments of translated texts to the written ones. How should we understand the arguments that some fragments are "great, funny, frightening, emotional, detailed or precise"? Did the translated fragments of scripts differ and how, since they did not arouse the same level of enthusiasm among the blind and partially sighted each time? Why do partially sighted people and those without vision impairment tend to prefer written scripts? And, finally, why did scene 2 arise enthusiasm of both the visually impaired and those without vision impairment?

The answers to the above questions can be probably found through a comparative analysis of the scripts created according to the strategy of writing and translating. Below you can find the results of such an analysis. In the first part, the research tool chosen for this analysis is presented, namely the cognitive linguistics. This fragment is followed by a detailed analysis of the scenes presented to the experimental group and the control group during the pilot study. Third part of this chapter includes the results of the cognitive comparative analysis of written and translated scripts, prepared for three films dubbed in Polish.

7.1 Selection of the research tool

Solving the question of time required recording the time spent on creating or translating the scripts. Getting to know the opinion of the audio description's target audience did not cause significant methodological problems as well. However, the comparison of the scripts in regard to their quality seems to be a Sisyphean task. How can we identify the differences between the scripts? How to make a comparison and which aspects should be compared?

One of the shortest, but, in this context, important and meaningful definitions of audio description says that it is a picture painted by words (see: 1.1). The available guidelines straightforwardly say: *describe what you see* (Audio Description Coalition 2009), which means that one has to describe what is visible in the picture, what can be observed there (Strzymiński & Szymańska 2010: 22). Seemingly, there is nothing easier than describing what you have before your eyes. It is commonly said that one picture is worth more than a thousand words. This popular saying sums up the audio describers' dilemmas very accurately. As it was mentioned before (see 2.1.3.) audio description has to fit in the intervals between the dialogues and should not conceal the important non-verbal elements of the soundtrack. With such a restricted time at hand, achieving a compromise between the aspects you would like to describe and the words that will fit between the dialogues is very difficult. The authors of standards and rules are obviously aware of the fact that it is impossible to describe everything. Therefore they come to rescue and give instructions to audio describers and teach them how to create audio description. The first rule, which has been mentioned already, is the rule of describing only important aspects (Strzymiński & Szymańska 2010: 20). It is also recommended that audio describers should answer the five wh-questions:

- **where:** description of the action's location and its changes
- **when:** information concerning the season, the time of a day, describing whether it is dark or bright, and even information concerning a given age
- **who:** physical appearance of the protagonists, including their age, characteristic features, clothes, color of skin or the way they move
- **what**: description of sounds that are hard to identify, subtitles belonging to the plot (e.g. headlines in a newspaper read by a protagonist), opening titles and end credits.
- **how:** action development: how the scene and the events are developed (López Vera 2006: 2).

The next recommendation is to choose **the most important elements** (Audio Description Coalition 2009), **important for the development of the action and the protagonists** (ADI AD Guidelines Committee 2003), **relevant for the plot** (Greening, Petré & Rai 2010) as well as valuable and necessary to reflect the aim and the character of a given production (Strzymiński & Szymańska 2010: 23).

These professional and specialist phrases seem to be nothing more than a euphemistic description of the fact that an audio describer is the one who subjectively selects some content from a picture. An individual decision process results in words and phrases, which are an intersemiotic translation of the picture. Therefore, in order to compare the scripts, it seems relevant to reverse the question repeated by audio describers and instead of asking "what" should be described and "how", one should ask "what" was described and "how" and in this way ask what was seen by the creators of particular scripts and what, as a consequence, was seen by visually impaired persons.

Until today, cognitive linguistics and its tools were not frequently applied to analyze audiovisual or intersemiotic translation. However, Agata Hołobut (2007) or Łukasz Bogucki and Mikołaj Deckert (2012) proved how useful the aforementioned tools can be in this field of translation studies. We should try to consider the reasons for using the tools of cognitive linguistics in the research on audio description.

In order to understand the difference between the scripts translated from English and those written in Polish, we would like to find an answer for the question of "what" and "how" was described in the scripts translated and written by different authors. Additionally, we gathered the opinions of the experimental group's members, who, by choosing the description of a given scene which was better in their opinion and explaining their decision, gave us hints concerning the nature of potential differences between scripts. Among the arguments given by the visually impaired in favour of a given description, the most frequent were those referring to the amount and details of the information included in it, as well as logicality or comprehensibility. At this point it seems necessary to explain what the respondents meant by stating that a given description included more information. When asked about this they explained that they meant those situations when a given description gave more information instead of concentrating on a detailed description of a give protagonist, piece of scenery or event. The above question and the hints given by the respondents seem to be inevitably compared to the concept of **construal** and the **dimensions of construal** in particular.

The crucial concept of cognitive semantics is that meaning consists of **conceptual content** and the way of construing the **content** (Langacker 2009: 70). **Imagery**, which is also often called **construal**, is connected with a basic human ability to portray the same situation in various ways in our mind (Taylor 2007:13). In different words, a given situation can be interpreted in many alternate ways, even if the conceptual content is the same (Langacker 1993: 19). The effects of this ability are reflected in a language – choosing particular phrases to describe a given situation is the consequence of the way the situation was construed in one's mind (Taylor 2007: 13). This question is investigated by Langacker (2009: 70) and exemplified by a glass containing water occupying half of its volume. He notices that at the conceptual level one can evoke this content in a neutral manner. However, the act of verbalisation imposes an image of the situation. The way the situation was construed in one's mind can be expressed by two different phrases: "The glass is half full", and "The glass is half empty". Choosing one expression instead of the other is not the matter of one's life optimism or pessimism, but rather a **designated relationship**. The first expression designates the relationship wherein the liquid occupies just half of the glass' volume. The second expression designates the relationship wherein the void occupies half of the glass' volume.

To go back to audio description and the question of comparing the scripts – it is hard to escape the impression that when asking "what" was described and "how", one asks in fact about the meaning understood as the resultant of the content and its construal, where the content is "what" was described and the construal is "how" the picture, the given state of affairs, is construed by a language user (Taylor 2007: 708).

When considering what audio describers chose from the entire picture they had access to, which in fact is the conceptual content they included in their descriptions, it is hard to resist the temptation of using the **visual metaphor**, in which conceptual content is compared to a **stage** and construal to a particular way of **viewing it** (Langacker 1995: 66). At this point one should emphasise that Langacker uses the word "stage" in a very particular way, and understands it as a construction which consists of a global **setting** – which does not participate in events and is only the place where they happen – and a determined number of smaller, more mobile **participants** who act and interact with one another (Langacker 2009: 471).

The starting point of the viewing metaphor presented in Picture 1 is visual perception.

Picture 1: *Viewing metaphor (Langacker 1995: 66).*

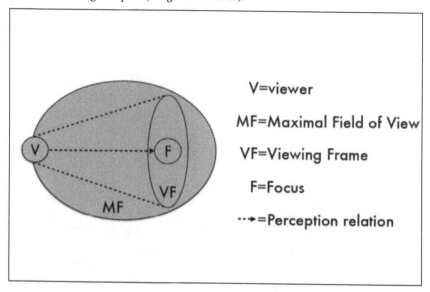

According to Langacker (1995: 66) a viewer has a given field of view, in other words – everything that we are able to see when we look. Viewing frame constitutes the area on which the viewer's attention is focused as the viewer distinguishes an object. This metaphor can be easily applied to audio description: the maximal field of view is constituted by everything that can be seen in a particular scene, and the viewing frame is the area selected and described by an audio describer, who particularly highlights a particular aspect.

According to the author himself, viewing metaphor suggests how to classify various dimensions of construal (Langacker 2009: 85). He explains that when viewing a scene, what we actually see depends on the distance from which we examine it, what we look at in particular, which objects we pay most attention to and where we view the scene from (Langacker 2009: 85). Each of the above mentioned parameters corresponds to the following dimensions of imagery: **specificity, focusing, prominence and perspective**. Subsequent parameters will be discussed on the basis of Langacker's typology (2009:85–128): level of specificity, scope, distinctness, iconicity and profiling.

A. *Level of specificity*

It is defined as the level of **precision and detail** in a description of a situation. Tabakowska (1995:62) compares levels of specificity to "fine" and "coarse" grain, since, as she writes, the scale determining the movement from a detailed picture to a scheme is obviously connected with the distance, from which we observe the scene. In other words, a close object is described more precisely and with more detail, while the same object seen from a bird's eye view is described in a schematic way.

B. Scope

Scope is the process of the initial selection of conceptual content to be portrayed in a language (Langacker 2008: 95). This aspect is closely bind to the level of specificity – following the concept of viewing metaphor one can say that the further the viewer is situated from the viewed object, the less precise and detailed is the viewing, but the wider the viewer's visual field (Tabakowska 1995:63).

C. Distinctness

Includes and connects two aspects, namely the above mentioned **scope** and the relationship between **figure and ground**. According to Langacker (2009: 99) all the selected things are distinguished as opposed to the things that are not distinguished, and the foreground is more important than the background. Tabakowska (1995: 67) argues that the **relation between figure and ground** is visible also on the higher levels of a conceptual structure, since human mind is able to organize conceptual content in such a way that one situation becomes a background to another one. Getting back to the viewing metaphor, Tabakowska (1995:68) explains that from the position of a viewer the foreground – which is the figure – is closer than the background – which is the ground – and this is why the observer pays more attention to the figure than to the ground. However, the observer can reverse this configuration and then the ground becomes more distinct.

D. Iconicity

Iconicity is understood as a direct similarity between conceptual structure and the form of linguistic expression (Tabakowska 1995: 79). At this point one should remember that a linguistic description, differently than a picture which as an effect of visual perception, is linear, which is imposed by linear nature of language (Tabakowska 1995: 79). Langacker points out that a speaking person usually reconstructs the sequence of perceiving events in a natural way (2009: 116). It is interesting that the same natural inclination is defined by Taylor (2007:57) as reconstruction of the sequence of real events' order and emphasizes that leaving an iconic order would demand clear emphasizing of the chronology of the events that happened.

E. Profiling

Expressions can profile both objects and relationships (Langacker 1995: 30–32). A profile's nature is closely bound to grammar categories. Nouns, pronouns and complex expressions profile **objects**; adjectives, adverbs, prepositions, infinitives and participles profile an **atemporal relation**, while personal forms of verb profile a **complex temporal relation**, which is called a process or an event (Langacker 1995: 32–34 and Langacker 2009: 151). The basis of the differentiation between objects, temporal and atemporal relations is the concept of time. Langacker differentiates **conceived time**, which can be briefly characterised as the time when an event happens, and **processing time**, which is understood as the time of conceptualisation (Langacker 2009: 152–153). Processes, which are temporal relations or events, are different from objects and atemporal relations since they are developed in the conceived time (Langacker 2009: 151).

Perceiving objects or relations assumes a mental operation called **scanning** (Langacker 2009: 152). Langacker differentiates the way of conceptualising objects and atemporal relations from temporal relations. The former are **scanned** in a **holistic way**, while the latter – in a **sequential** manner (Croft & Curse 2004: 53). The difference between sequential and holistic scanning is illustrated by Langacker through a reference to the difference between watching a film and looking at photographs (Evans & Green 2006: 535). **Sequential scanning** which happens in case of events is sometimes defined as an ability to observe a relation through time and compared to watching a movable film picture. Component states, which are the individual elements of the process, are activated through processing time in the order of their occurrence through conceived time in a non-cumulative way, which means that in a given processing moment only one component state is activated (Langacker 1995: 169 and Langacker 2009: 153). When it comes to **summary scanning** – as compared to looking at photographs – is such a kind of conceptualisation, in which, when scanning a complex scene and observing its elements, one sums up the elements that are perceived through individual stages (Langacker 2009: 121). As an effect, all the component states are available and active at the same time, and the constitute *gestalt* (Langacker 2009: 154). In summary, one can say after Langacker that thanks to scanning – in space in case of objects and through time when it comes to events – beings or component states are integrated to achieve constant concept of space or span of time (Langacker 2009: 152).

In this context, it is also worth analysing the aspect of the span of time or the temporal scope of events. **Temporal relations**, which Langacker (1995: 32–34) calls **processes**, can be divided internally into **perfective** and

imperfective processes, which are also called **immediate events – activities** and **extensive events – states.** They are profiled by perfective and imperfective verbs, which are also called **active** or **static** verbs (Langacker 2009: 198). According to Langacker the essence of the difference between imperfective ad perfective verbs is a time contour. Perfective verbs (e.g. *to fall*) designate a process that has both a beginning and an end, which is internally heterogeneous, which means that it covers the change that happens through time (Langacker 2009: 198). In other words, perfective processes are heterogeneous – every element of such a process is qualitatively different than the others, and when it is separated from the "parent" event, it becomes a distinct event (Tabakowska 2000: 25). On the other hand, imperfective verbs (e.g. *to have*) depict a process as a stable situation which continues without determined limits (Langacker 2009: 199). In contrast with heterogeneous perfective verbs, imperfective verbs are homogeneous – every component element of the process they profile is construed as identical (Langacker 2009: 199 and Tabakowska 2000: 25).

7.2 Research procedure

Firstly, a cognitive analysis of the chosen fragments of translated and written audio description scripts was conducted. The same fragments were presented to experimental group and control group during the pilot study. The aim of this analysis was to determine how the scene was constructed in the selected fragments. The results of this analysis, compiled with the preferences expressed by visually impaired participants of the pilot study, helped to deduce what type of a scene construction they prefer. In the analysis and conclusion only informative text function is taken into consideration, because this function was the argument which was most frequently expressed by visually impaired people in favour of one of the descriptions.

The next step was the analysis of audio description scripts prepared for films *Harry Potter and the Sorcerer's Stone, Harry Potter and the Prisoner of Azkaban* and *Ice Age 2: The Meltdown.* Three pairs of audio description scripts in total were subject to the analysis – two for each of the titles mentioned above. Every pair included both a written script and a script translated from English to Polish. Authors of the texts were novice audio desrcibers and novice translators. The aim of this analysis was to test how the authors of the scripts construct a scene and whether one of the ways dominates in a given author's scripts or in a given type of script, i.e. the written or the translated.

7.3 Detailed analysis

Below you can find a detailed cognitive analysis of the translated and the written texts, referring to each of the six scenes presented to both the experimental group and the control group in the pilot study. The scenes are described chronologically, starting with the one, whose translated script was preferred by the biggest number of visually impaired people (cf. 6.2.2.2).

A. Scene number five

The fifth scene presents the moment when Hagrid and Harry enter the pub called *The Leaky Cauldron*, and this is the first time when Harry experiences the magic reality of the world parallel to the world of muggles – people who do not have any magic skills and do not know that any world of magic exists.

Table 9: *Translated and written audio description of the fifth scene (backtranslation from Polish)*

Translated script	Written script
Hagrid is leading him to a bar called The Leaky Cauldron. The bar, filled with smoke, is illuminated by candles. Harry is looking at the customers who are wearing old-fashioned clothes. The bartender notices them.	Hagrid turns to the pub. He opens the door. The inside is dark and crowded. Pale candle lights glimmer. A beam of daily light falls into the room through one of the windows. Harry is looking around insecurely. Around him there are people dressed in old-fashioned coats and hats from more than two centuries before.

The descriptions of entering the bar differ considerably in both scripts. The only aspect which is the same in both scripts is iconicity, understood as preserving the sequence of the information given by the picture. In both scripts the sequence of shots was reflected. However, the differences are significant when it comes to scope, distinctness and level of specificity.

The authors of both scripts found the fact of entering the bar important, as well as the presence of candles and patrons in unusual clothes inside. However, it is worth considering that the translated text describes entering the bar with less detail than the written text, in which this activity is divided into two component processes. In the written script, Hagrid not only turns to the pub, but also opens the door. In this case the script is very iconic in relation to the picture – it accurately reflects what is shown by the camera in a given moment. When it comes to the translated script, its author not only assumed that the viewer will know that one should open the door in order to get inside, but also that he will recognise the door's sound when they are opened and the sound of their slamming.

It seems that thanks to resigning from such an iconic attitude to the description of the picture, the author saved space which is so valuable in audio description, and could give an information that the people entering the pub were noticed by a bar attender, and which information was not included in the written text. Importantly, this information is given even before the bar attender is visible on the screen. Therefore the translated script has a wider scope.

The differences between the scripts are mostly noticeable when it comes to the things each of the authors considered as distinct. In the translated scripts the action is distinct. The bar is described in only one sentence, which provides the information that the bar is filled with smoke and illuminated by candles. The information about the presence of unusual patrons is included in a sentence starting with a verb which draws attention to a process and not a relation. While it seems that the written script consider as distinct the decoration of the interior, which is described with many details. One can not only learn from the description that the bar is illuminated by candles, but also that it is dark, crowded and where the beam of light falls from.

To summarise, one can say that in the scene number five the translated audio description is characteristed by a smaller level of specificity and a broader scope than the written audio description. What is more, in the translated script action is more distinct, while in the written one the interior's description is more detailed. Both scripts are iconic, since they reflect the sequence of events represented in the picture, but the written script is more iconic when it comes to an accurate description of what is visible on the screen in a given moment.

In the case of this scene, as many as 92% of the visually impaired respondents preferred the translated audio description. The written script was chosen only by 8% of the survey's participants. The argument in favour of the translated script was that it is less detailed and more interesting. It is also worth noticing that in the case of this scene, the translated audio description was highly evaluated also by people withought vision impairment who emphasised that the translated script is more dynamic and interesting, although it describes the scenery in a less vibrant way.

B. Scene number six

The second scene, in which the translated audio description was preferred by the largest percent of visually impaired respondents, is one of the first scenes of the film. It shows the moment when Hagrid, driving a flying motorcycle, brings new-born Harry Potter to Privet Drive street, to place him in his relatives' care, together with professor McGonagall and professor Dumbledore.

Table 10: Translated and written audio description of the sixth scene (backtranslation from Polish)

Translated script	Written script
Bright light flashes over the trees. They are lamps of a motorcycle. Towering Hagrid lowers the motorcycle, lands and stops the machine with a squeal of tyres. He has a thick black beard and sparkling eyes which are barely visible from under his bushy eyebrows.	Professor McGonagall walks with the bearded man. They both turn around and notice a luminous ball in the sky that is heading towards them. The light turns out to be a front lamp of a motorcycle which lands right next to them. A huge, shaggy man in a long coat is sitting on the motorcycle. He turns off the motorcycle and lifts his goggles.

When comparing the description with the film, one can easily say that the written script is iconic, which means that it reflects the sequence of film shots accurately, while the text of the translated audio description goes beyond the order determined by the picture and "sits" Hagrid on a motorcycle before the giant is visible on the screen.

The differences are also visible when it comes to the scope. The written script mentioned four events, which were completely omitted in the translated script (and the original one): the description of walking wizards who notice a light, the fact of turning off the motorcycle and removing goggles by Hagrid.

Similarly to the previous scene, one can easily notice the difference concerning the aspects the authors assumed to be distinct, especially in the description of the flying motorcycle, which seems to be the key element of the scene. However, in the written script the "luminous ball" seems to be only in the background of the activities of professors McGonagall and Dumbledore, who is presented as an unidentified bearded man. On the other hand, in the translated script the information on the light that appears is the most distinct element of the scene.

It is also worth paying attention to the sentences about the motorcycle landing. In the written audio description script, the information on the nature of the light is more distinct than the fact of the landing itself. Which means that, · similarly to the previous scene, an element of scenery becomes more distinct than the action.

When it comes to the translated audio description script, in this scene it is less iconic in relation to the picture, it has a broader scope and the distinct element is the action.

The translated script of the scene number six was preferred by as many as 77% of the visually impaired respondents. The arguments in favour of this scene referred to the fact that this description is more comprehensive and accessible.

However, 56% of the people without visual impairment who chose the written script said that it is more accurate and authentic.

C. Scene number one

The fragments of the translated and written script presented below describe a particularly dynamic scene from the first moments of the film, when the letters from Hogwarts fall through a fireplace into the living room where Harry's foster family and he himself are sitting.

Table 11: *Translated and written audio description of the sixth scene (backtranslation from Polish)*

Translated script	Written script
A letter falls in through a fireplace. Uncle Vernon blinks surprised. Along with a strong gust of wind, a pile of new letters falls into the living room. Uncle Vernon plugs his ears. Frightened Dudley jumps onto his mother's lap. Harry hops up and down with delight. The letters fly in circles over their heads, as if in a snowstorm, and then fall on the floor to cover the carpet. Harry catches one of the letters and runs away to his cupboard.	A flying envelope hits Vernon's head. The whole family looks at the fireplace. Vernon grabs his head. An avalanche of letters falls into the floor through the fireplace. A hail of letters covers the whole room. Dudley finds a shelter on his mother's lap. Harry reaches out his hands and catches one of the letters.

Starting with iconicity, one can say that the written script is more iconic, both when it comes to preserving the sequence of events in the scene and the most faithful representation of what is visible on the screen in a given moment. In fact, the author of the written script uses the technique of summary scanning, by summing up individual shots, in which Vernon, Harry and Petunia with Dudley look at the fireplace, and by calling them "family". However, at the same time the moment when Harry catches a letter is described in a very iconic and detailed way, since this moment is divided into component parts of "reaching out his hand" and "catching the letter".

At the same time it is visible that the scope of the translated text is broader. The author of the written script passes over such happenings as the letter falling out of the fireplace or hopping with delight.

The way in which the falling in letters are described is also worth noticing. In the translated script a metaphor of snow is used, which is quite complex and sophisticated for such a short description: there is a snowstorm, the verb "fly" in circles (like snowflakes), an adverbial "over their heads" (again, like snow-flakes), a verb "fall" (see above) and an infinitive "to cover" (like snow covers the ground). This is a consistent, diversified metaphorical picture of a phenomenon – the letters are like a snowstorm in each of the above mentioned aspects. The written script also refers to a meteorological metaphor, but a different strategy is visible: nouns "avalanche" and "hail" and verbs "falls" and "covers" are used.

To sum up, one can say that in the case of the first scene the translated script is less iconic than the written one. The translated script also has a broader scope. In both scripts a metaphor was used to describe the letters that fall into the room, but in the translated audio description the metaphor refers to a more common set phrase.

In the case of this scene, the translated audio description was chosen by as many as 75% of the visually impaired respondents. The argument in favour of this scene was that it appeals to imagination in a better way and that it is more vivid.

D. Scene number four

The scripts of the subsequent scene describe the first night the new students of Hogwarts spend in the school.

Table 12: Translated and written audio description of the fourth scene (backtranslation from Polish)

Translated script	Written script
The same night in the boys' bedroom. The neatly folded uniforms lay on the chairs in front of the beds with four pillars. All the boys sleep. Except Harry. He is in his pyjamas and sits on the stony window sill with his knees under his chin.	Gryffindors sleep in their bedroom. Their blue outfits as well as their scarves and red and yellow striped ties lay on the chairs next to them, Harry in his pyjamas sits at the open window, through which the moonlight falls into the room.

The fourth scene is particularly interesting when it comes to iconicity. None of the descriptions respect the sequence of the shots visible on the screen and they both start with determining the locale instead of a description of the costumes. The written script begins with determining the locale, while also giving the information that the boys are asleep. The author probably assumed that the viewer will guess the remaining information, since usually people sleep at night. The translated script gives explicit information on the time of action in the first sentence and expects that the audience will guess that what people usually do in a bedroom at night is sleep.

The translated script is more detailed, since it precisely states that Harry does not sleep, as opposed to other boys. An exceptional element of the written script is a detailed description of clothes, which in the translated script are described as "uniforms". The level of specificity shows what the authors considered as distinct. In the case of the translated text the action is more distinct, while the written script focuses on a detailed description of the elements of the interior.

Generally speaking, both scripts are not iconic. Again, the translated script presents action as more distinct, and the written script focuses on an element of scenography, which is described with many details.

When it comes to these two descriptions, the translated one was chosen by 70% of the visually impaired respondents, and the written text was chosen by only 30%. The most frequent argument in favour of the translated script was the fact that it was less detailed, according to the respondents who preferred it.

E. Scene number three

Subsequent descriptions refer to a terrifying scene, which was preceded by an argument between Hermione, Harry and Ron: the girl decides to run away to a bathroom, but she does not know about the danger that is hidden there.

Table 13: *Translated and written audio description of the third scene (backtranslation from Polish)*

Translated script	Written script
Wiping her tears, Hermione leaves the toilet. She looks at two green legs, thick as tree trunks. She is frightened and looks up at a horribly green troll who obstructs the passage. His terrible head, which seems like a smashed green potato, touches the ceiling.	Hermione leaves the bathroom, wiping her tears. She turns her head up. A monstrous troll stands in front of her, with a sallow-green bulk, sticking out ears and a pimply bulging belly.

While analysing the descriptions of Hermione's encounter with the troll it is hard to say how the analysed audio description texts differ. Both scripts reconstruct the sequence of what is happening on the screen accurately. Each of them describes almost the same elements of the scene. The scopes of both descriptions differ insignificantly. Whereas in the first script Hermione "looks at two green legs" first, in the second one this topic is omitted and the girl stands eye to eye with the "monstrous troll" immediately. The translated script informs the viewer that the troll obstructs the passage, while the written script gives only information about the troll standing in front of the girl.

It may seem that an equally insignificant difference is related to distinctness – both descriptions favour the troll's physical appearance over a description of action. However, when analysing the construction of the sentences one can notice a subtle, but significant difference in this matter. In the first parts of the sentences from the translated script, which describe the troll's physical appearance, one can find active verbs that profile a perfective process. These verbs are included in the phrases: "she looks at" and "she is frightened and looks up". Since the verbs are placed in such a manner, the most distinct element in the sentences including a lot of information on the troll's physical appearance is in fact a perfective process. When it comes to the written script, the physical appearance of the beast seems to be the most distinctive element. A static verb "stands" was used in the sentence, and the specified characteristic features of the troll are enumerated one after another. The whole description makes a static impression, in contrast with the dynamic description from the translated script.

The scripts differ also when it comes to the specificity of the troll's physical appearance's description. Additional picturesque similes are used in the translated script, which informs the viewer that the troll has thick legs, is terrible and green, has a deformed head and is tall, since he touches the ceiling with his head. The written script provides less information, since it mentions a monstrous green animal, its sticking out ears and obesity. Therefore one can say that the translated script is more detailed than the written one.

In summary, both scripts are iconic and they differ insignificantly only when it comes to their scope, which is broader in the translated text, which is at the same time more specified.

Similarly to the previous scenes, the translated script emphasises action and the written script emphasis the elements of scenography.

As far as this scene is concerned, it is worth adding that 68 percent of the visually impaired respondents preferred the translated script, but the big number of 32% preferred the written script. When asked about their preferences, the respondents said that they liked the description of the troll – his head's description in particular – because they were able to imagine him quite well. However, the amount of details raised some controversies, since there were too many and it was difficult to understand the scene.

F. Scene number two

The last of the analysed scenes is very special – in this case the written script was preferred by the biggest number of visually impaired people (although it did not win completely with the translated one). In this scene the first lesson of flying a broom is described.

Table 14: *Translated and written audio description of the second scene(backtranslation from Polish)*

Translated script	Written script
The broom turns Neville around, bounces off the stony wall of the tower and flies down. Miss Hootch reaches draws her wand.	An out-of-control broom shoves Neville to the sides, overturns him and spins him around and finally bounces off Hogwarts walls. Neville dives and flying low he heads directly for the pupils. Professor Hootch draws her wand.

It is visible at first glance that – except from iconicity – the two descriptions of the same scene differ. Whereas their scope is almost the same, they differ significantly when it comes to distinctness and level of specificity.

Both scripts describe the same elements of the scene, although the scope of the written script is slightly broader, since it provides information that "Neville [...] heads directly for the pupils", which is not included in the translated script. In both scripts the action is the distinct element, but it is more distinct in the written script, since it is described with more details. While in the written text "broom shoves Neville to the sides, overturns him and spins him around" before he hits the wall of Hogwarts and dives, in the translated text the broom only "turns Neville around".

When it comes to iconicity, which is understood as the imitation of the sequence of the shots, the scripts do not differ – both the written and the translated script imitate the sequence of the shots adequately.

Therefore the descriptions of this scene do not differ as far as iconicity or scope is concerned. Both the written and the translated script assumed that action is the most distinct element, but the written script is more focused on action, since it thoroughly describes every component part of the process of Neville being turned around by the broom.

It is also worth to notice that, in case of this scene, the written script was preferred by as many as 44% of the respondents from the experimental group, who perceived it as funnier and the one that describes the scene better. Nevertheless, the remaining 56% of the research's participants said that in this case the translated audio description is easier to understand.

7.3.1 Results of the detailed analysis

The detailed analysis helped us to discover and understand how the translated and the written texts differ. It seems that they vary in the way the scenes are constructed when it comes to the level of specificity, scope, distinctness and iconicity. When analysing the descriptions we observed that the translated scripts are characterised by a narrower scope and level of specificity and are less iconic in relation to the picture than the written scripts. In the translated scripts perfective processes were distinct, while in the written scripts imperfective processes and relations were considered as distinctive.

On the basis of the conducted analysis and after comparing its results with visually impaired people's preferences and their opinions one can state that they prefer those descriptions, which, generally speaking, are characterised by both a narrower scope and a lower level of specificity. Among those preferred were also the fragments, which were less iconic in relation to the picture, both when it comes to preserving the sequence of the events in the picture, and an accurate description of what is visible on the screen at a given moment. The respondents perceived those descriptions as more dynamic, interesting and comprehensible.

The analysis also showed that the use of picturesque metaphors influences the preferences of visually impaired people. The scenes in which they were used were described by the respondents as more vivid and appealing to imagination. It seems that thanks to their introduction visually impaired persons are eager to accept even those descriptions which are not in accordance with their aforementioned preferences.

Similar situation takes place in relation to the scenes, whose descriptions were considered as funny by the respondents. In the case of the second and third scene, a relatively big percentage of the visually impaired respondents (accordingly: 44% and 32%) preferred the written script, although the descriptions of the scenes they mentioned had a broader scope and higher level of specificity than the translated script. Obviously, the majority of the respondents still preferred the translated scripts with a narrower scope and lower level of specificity, but it is visible that the disparities between the preferences of the respondents are bigger than in the other scenes. When taking into consideration the fact that the people who preferred the written script of this scene pointed out the comicality as the main reason for their preferences, one can say that the poetic function of audio description is also meaningful for visually impaired persons and in some cases using it can compensate for the fact that the audio description includes too many details.

7.4 Comprehensive analysis

All the pairs of scripts which constituted the corpus of the third experiment were analysed in the same way. The purpose of this anlysis was to verify how a scene is constructed when it comes to scope, level of specificity, distinctness and iconicity in translated and written scripts. The results of the comprehensive analysis helped us to evaluate whether one of this approaches prevails in the works of a given author or in a given type of script, i.e. the written or the translated one.

Generally speaking, the comprehensive analysis confirmed the results of the detailed analysis and proved that, indeed, the translated and written texts differ as far as the construal is concerned, in such aspects as scope, level of specificity, distinctness and iconicity.

However, the most important discovery of the comprehensive analysis is the question of iconicity, which was considered as barely significant at the beginning. Actually, after analysing all of the scripts one can say that iconicity is the element that influences and has an effect on the final shape of a script. Therefore the results of the comprehensive analysis will be presented from the perspectve of iconicity, which is understood as the will to reflect as accurately as possible the elements that are visible on the screen in a given moment, and somehow preserve the most accurate sequence of the description in relation to the sequence of the shots on the screen. To many audio desrcibers it is the indicator of "objectivity" and aplication of the rule "describe what you see".

During the analysis two types of attitude towards iconicity were observed. One can say that some of the texts moved simultaneously with the camera, and some of them did not follow the picture so thoroughly. The term "type of attitude" is used here, since it seems that the question of sctrict adherence to iconicity is rather the matter of an individual choice of the describer. However, it is worth noticing that, as it was proved by the analysis of the whole corpus, none of the attitudes is characteristic for a given author or a given strategy of creating text only. Both attitudes were visible in all of the scripts, but they differed in the matter of how frequently they are used.

A part of the descriptions is characterised with **more iconicity**, which is a bigger tendency to move simultaneously with the camera. The descriptions that move with the camera have a tendency to develop a higher **level of specificity**.

It is related with both taking into consideration particular component parts of a given process, and the elements of scenography. Examples of such an attachment to a picture as far as the description of a process is concerned can be constituted by the exaplmes preseneted in the detailed analysis part, which show the process of "reaching out hands and catching a letter" or "turning to the bar and opening the door". There are multiple similar examples in the whole corpus. However one of the most vivid examples of iconicity understood in this way, which is borderline absurd, comes from the *Harry Potter and the Sorcerer's Stone*.

It is a scene, in which Harry Potter stands in frot of a mirror and "lifts the corners of his mouth", instead of simply smiling. A similar tendency is visible in the description of scenography. Here, a perfect example is the description of clothes that lay on the chairs in the Gryffindors' bedroom. The script which is more iconic in this scene, describes individual parts of clothing in a great detail, while the second script presents them as "uniforms". There are consequences of such an attitude towards iconicity. As we know from the visual metaphor, the closer and observer gets to the scene, and the more detailed is the view, the smaller is the visual field. This rule is visible when it comes to the analysed scenes – the more details, the **narrower is the scope**.

It seems that iconicity influences also **distinctness**. The descriptions that follow the camera are constructing the scene "literally" and following the rule "here and now". Their descrptions are "faithful" to the shots, but not necessarily to the parts that are most distinctive in the whole scene or film.

A perfect example is the scene that describes how the protagonists enter the pub The Leaky Cauldron. When following the camera and the subsequent shots, the author of the description was focused on the description of the bar's interior. As a consequence, the scenery became the most distinctive element of this scene instead of the ongoing action. Therefore it seems that sometimes the authors who aim at iconicity choose the elements that are most distinctive in a given shot and not the most distinctive elements of a scene. In this way they create a translation which reflects the sequence of shots accurately, but is not functional.

There is also a question of changing the sequence of audio description in relation to the picture, which was observed in the descriptions of some scenes, e.g. in the description of the scene of landing in front of Dursley's house, in which one of the scripts provided information that Hagrid is a driver, even before he was visible on the screen. Probably the best example of such a procedure is a description of one of the first scenes from the film *Harry Potter and the Sorcerer's Stone*, which is presented below.

Table 15: The change of the sequence of description in relation to the picture (back translation from Polish)

Written script	Translated script
Harry gets up, turns the light on and puts on his glasses. Dudley – a corpulent peer of Harry – runs down the stairs. He turns back and stamps on the stairs. Harry's cupboard is under the stairs.	Harry Potter wakes up in his cupboard under the stairs and puts on his big, round glasses glued with a scotch tape. The corpulent boy runs down the stairs.

The written script reflects the sequence of shots accurately and the information on the place where Harry's cupboard is situated is given at the end of the scene's description. The less iconic translated script moves this information to the beginning of the description. Although the written text is iconic, one can have many objections concerning its consistency.

The analysis of all of the scripts proved that none of the types of attitudes towards iconicity is exclusive to either a given strategy of writing the scripts, or to particular authors. Both types of attitude towards iconicity can be noticed in all scripts, although with different frequency. The written scripts present iconicity more often. As a consequence, they often have narrower scopes, but include many more details. The analysis proved also that the scripts which are originally written in Polish include fewer metaphors and similes and, in contrast with the scripts translated from English, they considered scenography as distinctive more often than action.

7.5 Comparative analysis of the scripts – conclusions

On the basis of the detailed and comprehensive analysis, a conclusion can be drawn that visually impaired people seem to prefer those descriptions which are characterised by a narrower scope, lower level of specificity, focus their attention on action and are not iconic in relation to the picture. The research proved also that using metaphors and preserving the function of the text influences the preferences of visually impaired people. It can lead to a conclusion that when creating audio description one should take into consideration not only the informative function of the text, but also the poetic one, in particular when the audio description is created for a film which has a particular aim, like making the audience laugh or frightening them.

The detailed analysis suggested, and the comprehensive analysis confirmed the great importance of scripts' iconicity, referring directly to the general rule of creating audio description: "describe what you see". When speaking about iconicity we made a reference to the definition provided by Langacker (2008) and Taylor (2007). Probably both scientists meant the same when they said that an iconic description is the one that reflects the sequence of events. However, in the context of audio description and film, their words receive a new meaning. Taylor states that "when narrating a series of events, a speaker will strive to mention the events in the same order in which they occurred" (2007:54). The results of the research proved that in the context of audio description it is worth differentiating the sequence of presenting the events on screen and the usual or logical sequence of occurrences. A film, as a picture, is not linear, while audio description, as a description constructed by the means of a language, is linear. Visually impaired persons who participated in the research could choose among iconic descriptions, which followed the picture closely, but were formulated against linear logic, and they choose those descriptions which seemed to be more consistent as far as chronology, understood as a usual sequence of events, which one knows from experience, is concerned. Therefore, it seems that in case of writing audio description scripts the chronological or logical iconicity is more relevant than iconicity which is based on the picture. Obviously, we mean such films or even scenes, to be more precise, in which iconicity does not play an important role when it comes to the plot. *Memento* (dir. Christopher Nolan, 2000), a film which was "shot backwards", or films, in which the reversed iconicity can reveal the finale too early can be given as examples here.

As it was mentioned at the beginning, the analysis of the entire corpus proved that none of the attitudes towards iconicity is characteristic exclusively to a given author or a strategy of text creation. Both attitudes were visible in all of the scripts, but they differed in the matter of frequency. Comprehensive analysis showed also that in the translated scripts the attitude which can be called **logical iconicity** is more frequent. In most of the cases translated scripts have a narrower scope, lower level of specificity and action is their most distinctive element. Translated scripts are also the ones that include picturesque metaphors more often. Therefore, on the basis of the pilot study's and comparative cognitive analysis' results, it seems that translated scripts answer the needs and expectations of visually impaired people better. This is why one can state that, at least in case of the corpus we have analysed, visually impaired people who participated in this research would prefer translated scripts to written ones.

The use of chronologic iconicity or a poetic function seems to be opposed to the rules of creating audio description, which dictate to describe the things which are visible on the screen at a given moment, objectively and without interpretation. However, it seems that in case of audio description – which has to serve blind people by definition – the real needs and expectations of the target audience are more important than the rules.

III. Final Conclusions

What conclusions can be drawn from the three experiments which were conducted? Let us start with the one which seems to be the most important from the audio description's perspective – the target group's reaction to a proposed solution. On the basis of the research results we dare to say that the potential recipients of audio description, who are visually impaired people, would be eager to accept a script's translation from English to Polish, at least in case of dubbed films. What is more, it seems that due to greater experience of British audio describers, the scripts prepared by them are adapted to the needs of visually impaired persons better. The survey we have conducted proved that the translated audio description scripts were the ones preferred by more respondents, both the blind and partially sighted ones. Obviously, we are aware of the fact that our survey had limited scope and that its results can depend on the respondents' age and the genre of the film they were presented. Therefore, it seems inevitable to conduct a research on the reception of films with written and translated audio description, with a bigger and more diversified experimental group, and with other film genres. Due to the types of audiovisual translation which are used in Poland, it seems necessary to conduct a research on the possibility of translating audio description scripts to foreign films with Polish voice-over and audio subtitles.

Of course, we are aware of the fact that the research was conducted using some of the first scripts created in Poland, written in years 2006–2007, and scripts of novice audio describers. Undoubtedly, since that time Polish audio describers gained experience which helped them to create scripts which are better suited to the needs of the audience. We take into consideration the fact that the final shape of a script depends on individual skills and predispositions of an audio describer and the school he or she was trained in.[37] The arguments mentioned above makes us think that the scripts created as a result of the strategy of writing will not always be of lower quality than those which are the results of the strategy of translating. However, in the light of the research we have conducted, we can ascertain that the scripts created as a result of the strategy of translating can be at least equal in quality to those which are the results of the strategy of writing. Therefore, we suppose that in countries, in which audio description does not

---

37 There is an unofficial division into „Białystok" school, which is perceived as more strict when it comes to objectivity, and "Warsaw" school, which accepts poeticalness of a description.

have long tradition, the translated scripts can be of even higher quality than the written ones. This is why using the strategy of translation in such countries – and there are many of them – at least in its initial phase, can be beneficial for both the recipients of audio description and, for example, audiovisual translators, who have an opportunity to gain new skills and broaden their professional portfolio.

When it comes to time consumption of the processes of writing and translating a script, we proved that the latter is nearly three times faster, when the time of the work of novice audio describers and novice translators is compared. Interestingly, translation of a script from English to Polish, even when it is done by a novice translator, takes less time than writing a script not only when it is done by a novice audio describer, but also a person with more experience – in this case the time consumption difference equals two times.

It means that from the perspective of time consumption and, analogously, the costs, translating audio description scripts can be more beneficial. When facing these results one should also conduct a research on the time consumption of the strategy of translating scripts, which would be done by professional translators with great experience. It can be assumed that the difference between the time consumption of the strategy of translating and the strategy of writing would be even bigger.

People say that the biggest discoveries happen accidentally. While conducting the studies we, in a way incidentally, proved that the recently formulated questions and doubts concerning the usefulness of rules and standards of creating audio description are justified. The survey conducted among the visually impaired and the cognitive analysis proved that the audio description which strays from the rule of objectivity and describing the elements that are visible on the screen, and which is more narrative-like, fulfils its task better, and the task is to facilitate the reception of the works of visual and audiovisual culture to visually impaired people. However, it seems necessary to continue research in this field.

At the beginning of my PhD studies I had an opportunity to participate in a summer programme for young translation studies scholars, organised by the Centre for Translation Studies at the Catholic University in Leuven. We were asked to present papers on our research objectives. After each presentation professor Andrew Chesterman asked: "So what?" We understood quickly that this question was not a mockery of an eccentric professor, but a hint for future. Andrew Chesterman conveyed an important message hidden in those two words: the research we conduct should have a purpose.

As a result of the conducted research it was proved that we managed to prove that the audio description created by applying the strategy of translation is of good quality, is accepted by visually impaired people, and additionally is less time-consuming and cheaper. So what?

It seems that the results will be visible in five dimensions: scientific, market-related, educational, social and personal. **Scientific**, because those results are the first of their kind and open the doors for many research project, like a research on the possibilities of translating audio description scripts for the films with voice-over and audio subtitles; translating or creating audio description scripts for foreign films by translators who are more aware of cultural implications of a picture than audio describers who are not educated in translation studies; or a research on the possibilities of translating audio description scripts from English in contexts other than the Polish one. **Market-related**, because the creation of a tool that can decrease the time and cost, and at the same preserve high quality, was found interesting by distributors and broadcasters. **Educational**, because it seems that translation of audio description scripts is a perfect training for future audio describers, which allows them to learn what they should pay attention to when they will write scripts themselves. **Social,** since decreasing the time and cost of creating audio description can mean a greater availability of audiovisual products for the visually impaired. One can also consider applying this strategy to making other products of visual culture accessible, like descriptions of works of art or architecture. And, finally, I would like to write about the **personal** dimension, because the work on this book was an unusual adventure and a journey to the world of whose existence I was not aware before. Although I struggled with many obstacles, I would not hesitate to repeat this journey. In the course of this journey I met amazing friends and inspiring people, like blind Jan, who has travelled on his own for many years despite of his handicap and grey temples. Thanks to this journey the Seventh Sense Foundation was established. The Foundation implements science into people's life and thanks to that science does not serve for science only, but also for the good of a man.

References

Arma, S., 2011, *The Language of Filmic Audio Description: a Corpus-Based Analysis of Adjectives*, Universita degli studi di Napoli Federico II: Neapol.

Asimov, N., 2009, "August Coppola, arts educator, dies at 75", [access 27.12. 2012] http://www.sfgate.com/education/article/August-Coppola-arts-educa tor-dies-at-75-3282007.php.

Audio Description Coalition, 2009, *ADC Standards*, [access: 13.1.2012] http:// www.audiodescriptioncoalition.org/adc_standards_090615.pdf.

AVTLab., 2012, "Audiodeskrypcja z syntezą mowy", [access 20.1.2013] http://avt. ils.uw.edu.pl/ad-tts.

BBC. (b.d.). "British Broadcasting Corporation – Audiodescription on TV", [access: 20.3.2013] 20.02.2012] http://www.bbc.co.uk/aboutthebbc/ insidethebbc/howwework/policiesandguidelines/audiodescription.html.

Benecke, B., 2004, "Audio-Description", "Meta: Translators' Journal" 49/1: 78–80.

Benecke, B., 2007a, "Audio description: Phenomana of Information Sequencing", [access 6.1.2013] http://www.euroconferences.info/proceedings/2007_ Proceedings/ 2007_proceedings.html.

Benecke, B., 2007b, "Film AD", audio description workshop conducted during Advances Research Seminar in Audio Description, Universitat Autónoma de Barcelona: Barcelona.

Bogucki, Ł. and M. Deckert, 2012, "Kompetencja tłumaczeniowa a proces tworzenia interjęzykowych napisów filmowych", [in:] *Kompetencje tłumacza*, eds. Piotrowska, A. Czesak, A. T. Gomola, Kraków: Tertium, 107–124.

Bourne, J. and C. Jimenez Hurtado, 2007, "From the visual to the verbal in two languages: a contrastive analysis of the audio description of The Hours in English and Spanish", [in:] *Media for All: Subtitling for the Deaf, Audio Description, and Sign Language*, eds. J. Díaz Cintas, P. Orero, A. Remael, Amsterdam: Rodopi, 174–187.

Braun, S., 2008, "Audio description research: state of the art and beyond", "Translation Studies in the New Millenium", 6: 14–30.

Butkiewicz, U., Künstler I., Więckowski R. and A. Żórawska, 2012, *Audiodeskrypcja – zasady* tworzenia, [access 20.01.2013] http://dzieciom.pl/wp-content/ uploads/2012/09/Audiodeskrypcja-zasady-tworzenia.pdf.

Chancellery of the Sejm, *Ustawa z dnia 25 marca 2011 r. o zmianie ustawy o radiofonii i telewizji oraz niektórych innych ustaw,* [access: 10.7.2012] http://isap.sejm.gov.pl/DetailsServlet?id=WDU20110850459.

Chaume, F., 2004, "Synchronization in dubbing: A tranlational approach", [in:] *Topics in Audiovisual Translation,* eds. P. Orero, Amsterdam: John Benjamins, 35–52.

Chmiel, A. and I. Mazur, 2011, "Overcoming barriers – The pioneering years of audio description in Poland", [in:] *Audiovisual Translation in Close-up,* ed. W. A. Serban, A. Matamala, J.-M. Lavaur, Bern: Peter Lang, 279–296.

Ciborowski, M., 2010, "Pod lupą: Chodź, opowiedz mi świat", [access 20.7.2012] http://pochodnia.pzn.org.pl/artykul/311-pod_lupa_chodz_opowiedz_mi_swiat.html.

Croft, W. and D.A. Curse, 2004, *Cognitive Linguistics.* Cambridge: Cambridge University Press.

Cronin B., Evans D. and M. Pfanstiehl, 1997, *Report of The National Coallition of Blind and Vusually Impaired Persons for Increased Video Access,* [access 24.1.2013] http://ecfsdocs.fcc.gov/filings/1997/07/22/187612.html.

Delabastita, D., 1989, "Translation and mass-communication: Film and T.V. translation as evidence of cultural dynamics", "Babel International Journal of Translation", 4: 193–218.

Díaz Cintas, J., 2007, "Por una preparación de calidad en accesibilidad audiovisual", [in:] *Trans* 11: 45–60.

European Parliament and Council, 2007, *Directive 2007/65/WE of the European Parliament and Council,* [access 5.5.2012] http://eur-lex.europa.eu/LexUriServ/LexUriServ.do?uri=OJ:L:2007:332:0027:0045:PL:PDF.

Evans, E. and R. Pearson, 2009, "Boxed out: Visually Impaired Audiences, Audio Description and the Cultural Value of the Television Image", "Participations – Journal of Audience & Reception Studies", 6/2: 373–402.

Evans, V. and M. Green, 2006, *Cognitive Linguistics. An Introduction.* Edinburgh: Edinburgh University Press.

Farnham, A., 2000, *Forbes great success stories: twelve tales of victory wrested from defeat.* New York: John Wiley and Sons.

Frazier, G., 1996, *Report no. mm 95–115,* [access 27.12. 2012] http://ecfsdocs.fcc.gov/filings/1996/01/25/154152.html.

Fryer, L., 2009, *Directing in Reverse*. [access 5.2.2013] http://www.port.ac.uk/departments/academic/slas/conferences/pastconferenceproceedings/translationconf2009/translationconf2009file/filetodownload,138144,en.pdf.

Gambier, Y., 2003, "Screen Transadaptaion: Perception and Reception", "The Translator: Special Issue on Screen Translation", 9/2: 171–189.

Georgakopoulou, Y., 2009, "Developing audio description in Greece", "MultiLingual", 38–42.

Gerzymisch-Arbogast, H., 2007, "Workshop Audio Description", [acces 5.12.2012] http://www.translationconcepts.org/pdf/audiodescription_forli.pdf.

Greening, J., 2010, interview with Anna Jankowska, Berlin.

Greening, J. and D. Rolph, 2007, "Accessibility: raising awareness of audio description in the UK", [in:] *Media for All: Subtitling for the Deaf, Audio Description, and Sign Language*, eds. J. Díaz Cintas, P. Orero, A. Remael, Amsterdam: Rodopi, 127–138.

Grodecka, A., 2010, "Ekfraza w edukacji", [access 1. Downloaded on 10.10.2012] http://www.staff.amu.edu.pl/~grodecka/ekfraza/ekfraza%20w%20edukacji.pdf.

Hernàndez- Bartolomé, A. o G. Mendiluce Cabrera, 2009, "How can images be translated? Audio description, a challenging audiovisual and social gap filler", [in:] "Hermeneus" 11: 161–186.

Herrador Molina, D., 2006, *La traducción de guiones de audiodescripción del inglés al español*, Granada: Universidad de Granada.

Hołobut, A., 2007, *Projekt przedmiotu użytkowego a jego obraz językowy*, Kraków: Uniwersytet Jagielloński.

Hopfinger, M., 1974, *Adaptacje filmowe utworów literackich. Problemy teorii i interpretacji*, Wrocław: Zakład Narodowy im. Ossolińskich.

Hyks, V., 2005, "Audio Description and Translation. Two related but different skills", "*Translating Today*", 4: 6–8.

Igareda, P., 2011, "The audio description of emotions and gestures in Spanish-spoken films", [in:] *Audiovisual Translation in Close-up*, eds. W. A. Serban, A. Matamala, J.-M. Lavaur, Bern: Peter Lang, 223–238.

Independent Television Comission, 2000, *Guidance on Standards for Audio Description*, [access 10.08.2012] http://www.ofcom.org.uk/static/archive/itc/itc_publications/codes_guidance/audio_description/audio_1.asp.html.

Institute of Informatics of the Jagiellonian University, 2002, "Działalność Pracowni Tyfloinformatycznej", [access 30.12.2011] http://www.ii.uj.edu.pl/archiwum/Dzienne_tyflo.html.

Jakobson, R., 2012 [1959], "On linguistic aspects of translation", [in:] *The Translation Studies Reader*, ed. L. Venutti, Nowy Jork: Routledge, 126–132.

Jakubowski, S., 2007, "Filmy dla niewidomych", "Nowy Magazyn Muzyczny" [access 30.12.2012] http://www.idn.org.pl/towmuz/artykuly1.html#21a.

Jankowska, A., 2008, "Audiodeskrypcja – wzniosły cel w tłumaczeniu", [in:] *Między oryginałem a przekładem: wzniosłość i styl wysoki w przekładzie*, eds. J. Brzozowski, M. Filipowicz-Rudek, Kraków: Księgarnia Akademicka, 225–256.

Jankowska, A., 2010, "Translating vs. Creating: Developing Audio Description in Poland". referat wygłoszony podczas konferencji Languages and the Media, Berlin.

Jankowska, A., 2012, "Kompetencje tłumacza audiowizualnego" [in:] *Kompetencje tłumacza*, eds. M. Piotrowska, A. Czesak, A. Gomola, S. Tyupa, Kraków: Tertium.

Jankowska, A. & A. Szarkowska, 2011, "Audiodeskrypcja – nowe trendy", [access: 4.1.2013] http://www.ckwz.art.pl/Anna_Jankowska-Audiodeskrypcja-nowe_trendy.pdf.

Jankowska A., M. Wilgucka & A. Szarkowska, 2014, „Audio Description for Voiced-over Films: the Case Study of Big Fish" in *Między Oryginałem a Przekładem 23*, 81–96.

Kalbarczyk, M., 2004, *Świat otwarty dla niewidomych. Szanse i możliwości.* Warszawa: WSiP.

Künstler, I., 2007, "Udostępniania niewidomym filmów – systemy audio deskrypcji i osiągnięcia w tej dziedzinie w krajach Unii Europejskiej i Ameryce Północnej", 5th Edition of the Conference: *Reha for the Blind in Poland. Miejsce inwalidów wzroku w społeczeństwie informacyjnym – zastosowanie technologii tyfloinformatycznej dla zniwelowania skutków inwalidztwa wzroku*, Warszawa: Unia Pomocy Niepełnosprawnym – Szansa, 56–59.

Langacker, R. 1995, *Wykłady z gramatyki kognitywnej*, Lublin: Wydawnictwo UMCS.

Langacker, R., 2009, Cognitive Grammar, Kraków: Universitas.

Link, K., 2002, "Melpomena z białą laską", [access 12.11.2012] ftp://superfon.my-ftp.org/prasa/archiwum%20biuletynu%20informacyjnego/2002/biuletyn.05/bi05_02.txt.

Lipińska, E., 2006, "Czynniki wpływające na proces uczenia się", [in:] Z zagadnień dydaktyki języka polskiego jako obcego, eds. E. Lipińska and A. Seretny Kraków: Universitas, 57–77.

López Vera, J. F., 2006, "Trasnlating audio description scripts – the way forward? Tentative first stage project results", [access 22.07.2012] http://www.euroconferences.info/proceedings/2006_Proceedings/2006_Lopez_Vera_Juan_Fran cisco.pdf.

Malzer-Semlinger, N., 2012, "Narration or description: What should audio description "look" like?", [in:] Emerging Topics in Audiovisual Translation: Audio description, ed. E. Perego, Trieste: EUT Edizioni Università di Trieste, 29–36.

Matamala, A. and P. Orero, 2008, "Designing a Course on Audio Description and Defining the Main Competences of the Future Professional", [in:] "Linguistica Antverpiensa" 6: 329–344.

Matamala, A., 2006, "La accesibilidad en los medios: aspectos lingüísticos y retos de formación", [in:] Sociedad, integración y televisión en España, eds. R. Amat and A. Pérez-Ugena, Madrid: Laberinto: 293–306.

Mazur, I. and A. Chmiel, 2011, "Odzwierciedlenie percepcji osób widzących w opisie dla osób niewidomych. Badania okulograficzne nad audiodeskrypcją", [in:]. "Lingwistyka Stosowana" 4.

Mazur, I. and A. Chmiel, 2012, "Towards common European audio description guidelines: results of the Pear Tree Project", [in:] "Perspectives: Studies in Translatology. Special Issue: Pear Stories and Audio Description: Language, Perception and Cognition across Cultures" 20/1: 5–23.

MoPix, date not provided, "MoPix Motion Picture Access" [access 12.01.2013] http://ncam.wgbh.org/mopix/.

Navarrete, F., 1997, "Sistema AUDESC: el arte de hablar en imágenes". [in:]"Integración" 23: 70–75.

Office of Communications, 2006, "Guidelines on the provision of television access service", [access: 2.2.2012] http://stakeholders.ofcom.org.uk/binaries/broadcast/guidance/guidelines.pdf.

Office of Communications, 2012, "Television Access Services: Full Year Cumulative Report 2011", [access 15.03.2013] http://stakeholders.ofcom.org.uk/market-data-research/market-data/tv-sector-data/tv-access-services-reports/

2011-cumulative?utm_source=updates&utm_medium=email&utm_campaign=access-services-2011.

Orero, P., 2005, "Audio Description: Professional Recognition, Practice and Standards in Spain", "Translation Watch Quaterly", 7–18.

Orero, P., 2007a, "Pioneering audio description: an interview with Jorge Arandes", "Jostrans Journal of Specialised Translation", 7: 179–189.

Orero, P., 2007b, "Sampling audio description in Europe, [in:] *Media for All: Subtitling for the Deaf, Audio Description, and Sign Language*, eds. J. Díaz Cintas, P. Orero, A. Remael, Amsterdam: Rodopi, 111–126.

Orero, P., 2012, "Film reading for writing audio descriptions: A word is worth a thousand images?", [in:] *Emerging Topics in Audiovisual Translation: Audio description*, ed. E. Perego, Trieste: EUT Edizioni Università di Trieste, 29–36. 13–28.

Pearson, R., and E. Evans, E., 2008, "Royal National Institute of the Blind: Boxed Out – Television and People with Sight Problems. Executive Summary", [access 22.10.2012] www.rnib.or.uk/xpedio/groups/public/documents/publicwebsite/public_adboxoutec.doc

Psiuk, A., 2010, *Analiza przedstawienia bohaterów w skrypcie audiodeskrypcji tworzonej i tłumaczonej do filmu Harry Potter i więzień Azkabanu, a strategie tworzenia audiodeskrypcji dola filmów dubbingowanych*, Kraków: Uniwersytet Jagielloński.

Rai, S., Greening, J., and L. Petre, 2010, "A Comparative Study of Audio Description Guidelines Prevalent in Different Countires", [access 20.4.2012] http://www.rnib.org.uk/professionals/solutionsforbusiness/tvradiofilm/Pages/international_AD_guidelines.aspx.

Remael, A. and G. Vercauteren, 2007, "Audio describing the exposition phase of films. Teaching students what to choose", "TRANS. Revista de Traductología", 11: 73–93.

Royal National Institute of the Blind, 2013, "Audio description on *DVD*", [access: 20.03.2013] http://www.rnib.org.uk/livingwithsightloss/tvradiofilm/film/Pages/dvd.aspx.

Seventh Sense Foundation, 2010, *Wytyczne dla organizatorów festiwali filmowych*.

Snyder, J., 2007, "Audio Description: The Visual Made Verbal", "The International Journal of the Arts in Society", 2/2: 99–104.

Snyder, J., 2008, "Audio description: The visual made verbal", [in:] *The Didactics of Audiovisual Translation*, ed. J. Díaz Cintas, Amsterdam: John Benjamins Publishing, 191–198.

Strzymiński, T. and B. Szymańska, B., 2010, *Standardy Tworzenia Audiodeskrypcji*, [access 3.09.2012] http://www.audiodeskrypcja.org.pl/index.php/standardy-tworzenia-audiodeskrypcji/do-produkcji-audiowizualnych.

Surdyk, A., 2006, "Metodologia action research i techniki komunikacyjne w glottodydaktyce", [in:] *Oblicza komunikacji 1: Perspektywy badań nad tekstem, dyskursem i komunikacją*, eds. I. Kamińska-Szmaj, T. Piekot and M. Zaśko-Zielińska, Kraków: Tertium, 912–923.

Szarkowska, A., 2008, "Przekład audiowizualny w Polsce – perspektywy i wyzwania", "Przekładaniec" 20: 8–25.

Szarkowska, A., 2011, "Text-to-speech audio description. Towards a wider availability of AD", "Jostrans Journal of Specialised Translation" 15: 142–162.

Szarkowska, A., 2013, "Auteur description – from the director's creative vision to audio description", [access 23.03.2013] http://avt.ils.uw.edu.pl/files/2013/03/Szarkowska_Auteur-description.pdf.

Szarkowska, A. & A. Jankowska, A., 2012, "Text-to-speech audio description of voiced-over films. A case study of audio described Volver in Polish", [in:] *Emerging Topics in Audiovisual Translation: Audio descriptio*, ed. E. Perego, Trieste: EUT Edizioni Università di Trieste, 29–36. 81–98.

Szczepański, H., 2001, "Co to jest tyflofilm?", [access 30.11.2012] http://www.opoka.org.pl/biblioteka/I/IC/tyflofilm.html.

Tabakowska, E., 1995, *Gramatyka i obrazowanie: Wprowadzenie do językoznawstwa kognitywnego*, Kraków: PAN.

Tabakowska, E., 2000, "Struktura wydarzenia w literackim tekście narracyjnym", [in:] *Przekładając nieprzekładalne*, ed. M. Ogonowska, Sopot: Wydawnictwo Uniwersytetu Gdańskiego, 19–37.

Taylor, J. R., 2007, *Gramatyka kognitywna*. Kraków: Universitas.

Thomas, R., 1996, "Gregory T. Frazier, 58; Helped Blind See Movies With Their Ears", [access 26.07.2012] http://www.nytimes.com/1996/07/17/us/gregory-t-frazier-58-helped-blind-see-movies-with-their-ears.html.

TNS OBOP, 2011, *TNS OBOP – Podsumowanie* widowni, [access 31.12.2011] http://www.obop.pl/telemetria/wyniki_2011_-_podsumowanie_tygodnia/tydzie_2011-11-21.

Tomaszkiewicz, T., 2006, *Przekład audiowizualny.* Warszawa: Wydawnictwo Naukowe PWN.

TVP, 2011, "TVP jako pierwsza stacja w Polsce wyemitowała film z audiodeskrypcją dla niewidomych", [access 02.03.2012] http://www.tvp.pl/o-tvp/centrum-prasowe/komunikaty-prasowe/tvp-jako-pierwsza-stacja-w-polsce-wyemitowala-film-z-audiodeskrypcja-dla-niewidomych/4698474.

Udo, J., 2009, interview with Anna Jankowska, Barcelona.

Udo, J.P., and *D. Fels*, 2011, "From the describer's mouth: Reflections on creating unconventional audio description for live theatre", [in:] *Audiovisual Translation in Close-up*, eds. W A. Serban, A. Matamala, J.-M. Lavaur, Bern: Peter Lang, 257–278.

United Nations, 2006, *The Convention on the Rights of Persons with Disabilities.* access 17.1.2012] http://www.unic.un.org.pl/dokumenty/Konwencja_Praw_Osob_Niepelnosprawnych.pdf.

Vandaelea, J., 2012, "What meets the eye. Cognitive narratology for audio description", "Perspectives: Studies in Translatology. Special Issue: Pear Stories and Audio Description: Language, Perception and Cognition across Cultures", 20/1: 87–102.

Washington Ear, 2002, "The Metropolitan Washington Ear – What Is It", [access 26.12.2012] www.washear.org/whatitis.htm.

Wilgucka, M., 2012, *Audio description for voiced-over films: the case study of Big Fish.* Kraków: Uniwersytet Jagielloński.

Woch, A., 2009, interview with Anna Jankowska, Kraków.

Appendix 1

Questionnaire for visually impaired people

I. GENERAL QUESTIONS

1. **Age**

   - ☐ Under 6
   - ☐ 6–12
   - ☐ 12–18
   - ☐ 18–35
   - ☐ 35–50
   - ☐ 50–65
   - ☐ Over 65

2. **Sex**

   - ☐ Male
   - ☐ Female

3. **Education**

   - ☐ Primary
   - ☐ Secondary
   - ☐ College/University degree

4. **Which of the sentences describe your sight (in glasses or contact lenses if you usually use them)? Imagine that you are in a well-illuminated room and answer "yes", "no" or "I don't know" next to each of the sentences. Do you see well enough to:**

   - ☐ Recognise the location of a window by light? ☐ Yes ☐ No ☐ I don't know
   - ☐ See the shapes of furniture in a room? ☐ Yes ☐ No ☐ I don't know
   - ☐ Recognise a person you know when you are close to his/her face? ☐ Yes ☐ No ☐ I don't know
   - ☐ Recognise a person you know when he/she stands within arm's reach? ☐ Yes ☐ No ☐ I don't know
   - ☐ Recognise a person you know when he/she stands in the opposite part of the room? ☐ Yes ☐ No ☐ I don't know

☐	Recognise a person you know when he/she stands on the other side of the street?	☐ Yes ☐ No ☐ I don't know
☐	Read a regular printed newspaper?	☐ Yes ☐ No ☐ I don't know
☐	Read a book with enlarged print?	☐ Yes ☐ No ☐ I don't know
☐	Read a headline of a newspaper?	☐ Yes ☐ No ☐ I don't know

5. At what age did you start losing your sight?

- ☐ I am blind since I was born
- ☐ 3 or under
- ☐ 4–18
- ☐ 19–59
- ☐ Over 60

6. How much time do you spend watching TV/DVD films daily?

- ☐ Less than 1 hour
- ☐ 1–2 hours
- ☐ 2–3 hours
- ☐ More than 3 hours

7. What kind of films and shows do you watch?

- ☐ Polish films and shows
- ☐ Dubbed foreign films and shows
- ☐ Foreign films and shows with a voice over
- ☐ Foreign films and shows with subtitles read by a speech synthesizer

8. Are you disturbed by foreign language voices you hear in the background of the films with a voice over?

- ☐ Yes
- ☐ No
- ☐ I don't know

9. What kind of audio described films would you like to watch the most?

- ☐ Polish ones
- ☐ Foreign ones
- ☐ I don't have an opinion

10. **Have you ever watched any audio described film?**

☐ Yes
☐ No
☐ I don't know

11. **If yes, how many hours of an AD film have you watched?**

Less than 5 hours
5–10 hours
10–20 hours
More than 20 hours

II. DETAILED QUESTIONS

12. **Have you noticed any difference between the audio description in the first and the second half of the film?**

☐ Yes
☐ No

13. **If you answered yes in question no 18, try to define how the two halves differed.**

14. **Which description of one of the scenes of the film do you prefer? Why?**

1) A letter falls in through a fireplace. Uncle Vernon blinks surprised. Along with a strong gust of wind, a pile of new letters falls into the living room. Uncle Vernon plugs his ears. Frightened Dudley jumps onto his mother's lap. Harry hops up and down with delight. The letters fly in circles over their heads, as if in a snowstorm, and then fall on the floor to cover the carpet. Harry catches one of the letters and runs away to his cupboard.

2) A flying envelope hits Vernon's head. The whole family looks at the fireplace. Vernon grabs his head. An avalanche of letters falls onto the floor through the fireplace. A hail of letters covers the whole room. Dudley finds a shelter in his mother's lap. Harry reaches out his hands and catches one of the letters.

15. **Which of the descriptions do you prefer?**

1) An out-of-control broom shoves Neville to the sides, overturns him and spins him around and finally bounces off Hogwarts walls. Neville dives and flying low he heads directly for the pupils. Professor Hootch draws her wand.

2) The broom turns Neville around, bounces off the stony wall of the tower and flies down. Miss Hootch draws her wand.

16. Which of the descriptions do you prefer?

1) Wiping her tears, Hermione leaves the toilet. She looks at two green legs, thick as tree trunks. She is frightened and looks up at a horribly green troll who obstructs the passage. His terrible head, which seems like a smashed green potato, touches the ceiling.

2) Hermione leaves the bathroom, wiping her tears. She turns her head up. She sees a monstrous troll with a sallow-green bulk, sticking out ears and a pimply bulging belly.

17. Which of the descriptions do you prefer?

1) Gryffindors sleep in their bedroom. Their blue outfits as well as their scarves and red and yellow striped ties lay on the chairs next to them, Harry in his pyjamas sits at the open window, through which the moonlight falls into the room.

2) The same night in the boys' bedroom. The neatly folded uniforms lay on the chairs in front of the beds with four pillars. All the boys sleep. Except Harry. He is in his pyjamas and sits on the stony window sill with his knees under his chin.

18. Which of the descriptions do you prefer?

1) Hagrid turns to the pub. He opens the door. The inside is dark and crowded. Pale candle lights glimmer. A beam of daily light falls into the room through one of the windows. Harry is looking around insecurely. Around him there are people dressed in old-fashioned coats and hats from more than two centuries before.

2) Hagrid is leading him to a bar called The Leaky Cauldron. The bar, filled with smoke, is illuminated by candles. Harry is looking at the customers who are wearing old-fashioned clothes. The bartender notices them.

19. Which of the descriptions do you prefer?

1) They both turn around and notice a luminous ball in the sky that is heading towards them. The light turns out to be a front lamp of a motorcycle which lands right next to them. A huge, shaggy man in a long coat is sitting on the motorcycle. He turns off the motorcycle and lifts his goggles.

2) Bright light flashes over the trees. They are lamps of a motorcycle. Towering Hagrid lowers the motorcycle, lands and stops the machine with a squeal of tyres. He has a thick black beard and sparkling eyes which are barely visible from under his bushy eyebrows.

Appendix 2

Questionnaire for people without sight malfunction

1. **Age**
   - ☐ Under 6
   - ☐ 6–12
   - ☐ 12–18
   - ☐ 18–35
   - ☐ 35–50
   - ☐ 50–65
   - ☐ Over 65

2. **Sex**
   - ☐ Female
   - ☐ Male

3. **Which description of one of the scenes of the film do you prefer? Why?**

   1) A letter falls in through a fireplace. Uncle Vernon blinks surprised. Along with a strong gust of wind, a pile of new letters falls into the living room. Uncle Vernon plugs his ears. Frightened Dudley jumps onto his mother's lap. Harry hops up and down with delight. The letters fly in circles over their heads, as if in a snowstorm, and then fall on the floor to cover the carpet. Harry catches one of the letters and runs away to his cupboard.

   2) A flying envelope hits Vernon's head. The whole family looks at the fireplace. Vernon grabs his head. An avalanche of letters falls into the floor through the fireplace. A hail of letters covers the whole room. Dudley finds a shelter on his mother's lap. Harry reaches out his hands and catches one of the letters.

4. **Which of the descriptions do you prefer?**

   1) An out-of-control broom shoves Neville to the sides, overturns him and spins him around and finally bounces off Hogwart's wall. Neville dives and flying low he heads directly for the pupils. Professor Hootch reaches draws her wand.

   2) The broom turns Neville around, bounces off the stony wall of the tower and flies down. Miss Hootch draws her wand.

5. **Which of the descriptions do you prefer?**

   1) Wiping her tears, Hermione leaves the toilet. She looks at two green legs, thick as tree trunks. She is frightened and looks up at a horribly green troll who obstructs the passage. His terrible head, which seems like a smashed green potato, touches the ceiling.
   2) Hermione leaves the bathroom, wiping her tears. She turns her head up. She sees a monstrous troll with a sallow-green bulk, sticking out ears and a pimply bulging belly.

6. **Which of the descriptions do you prefer?**

   1) Gryffindors sleep in their bedroom. Their blue outfits as well as their scarves and red and yellow striped ties lay on the chairs next to them, Harry in his pyjamas sits at the open window, through which the moonlight falls into the room.
   2) The same night in the boys' bedroom. The neatly folded uniforms lay on the chairs in front of the beds with four pillars. All the boys sleep. Except Harry. He is in his pyjamas and sits on the stony window sill with his knees under his chin.

7. **Which of the descriptions do you prefer?**

   1) Hagrid turns to the pub. He opens the door. The inside is dark and crowded. Pale candle lights glimmer. A beam of daily light falls into the room through one of the windows. Harry is looking around insecurely. Around him there are people dressed in old-fashioned coats and hats from more than two centuries before.
   2) Hagrid is leading him to a bar called The Leaky Cauldron. The bar, filled with smoke, is illuminated by candles. Harry is looking at the customers who are wearing old-fashioned clothes. The bartender notices them.

8. **Which of the descriptions do you prefer?**

   1) They both turn around and notice a luminous ball in the sky that is heading towards them. The light turns out to be a front lamp of a motorcycle which lands right next to them. A huge, shaggy man in a long coat is sitting on the motorcycle. He turns off the motorcycle and lifts his goggles.
   2) Bright light flashes over the trees. They are lamps of a motorcycle. Towering Hagrid lowers the motorcycle, lands and stops the machine with a squeal of tyres. He has a thick black beard and sparkling eyes which are barely visible from under his bushy eyebrows.

Text–Meaning–Context
Cracow Studies in English Language, Literature and Culture

Edited by Elżbieta Chrzanowska-Kluczewska and Władysław Witalisz

Vol. 1 Alicja Witalisz (ed.): Migration, Narration, Communication. Cultural Exchanges in a Globalised World. 2011.

Vol. 2 Bożena Kucała: Intertextual Dialogue with the Victorian Past in the Contemporary Novel. 2012.

Vol. 3 Cinta Zunino: *Mimesis* and the Representation of Experience: Dramatic Theory and Practice in pre-Shakespearean Comedy (1560–1590). 2012.

Vol. 4 Grzegorz Szpila: Idioms in Salman Rushdie´s Novels. A Phraseo-Stylistic Approach. 2012.

Vol. 5 Poetry and its Language. *Papers in Honour of Teresa Bela.* Edited by Marta Gibińska and Władysław Witalisz. 2012.

Vol. 6 Peter Leese/Carly McLaughlin/Władysław Witalisz (eds.): Migration, Narration, Identity. Cross-Cultural Perspectives. 2012.

Vol. 7 Marta Dąbrowska: Variation in Language. Faces of Facebook English. 2013.

Vol. 8 From Sound to Meaning in Context. Studies in Honour of Piotr Ruszkiewicz. Edited by Alicja Witalisz. 2014.

Vol. 9 Anna Walczuk/Władysław Witalisz (eds.): Old Challenges and New Horizons in English and American Studies. 2014.

Vol. 10 Bożena Kucała/Robert Kusek (eds.): Travelling Texts: J.M. Coetzee and Other Writers. 2014.

Vol. 11 Anna Jankowska: Translating Audio Description Scripts. Translation as a New Strategy of Creating Audio Description. Translated by Anna Mrzygłodzka and Anna Chociej. 2015.

www.peterlang.com